NATURAL AND SYNTHETIC SWEET SUBSTANCES

ELLIS HORWOOD BOOKS IN ORGANIC CHEMISTRY
Series Editor: Dr J. MELLOR, University of Southampton

Belen'kii, L.I.	**CHEMISTRY OF ORGANOSULFUR COMPOUNDS: General Problems**
Beyer, H. & Walter, W.	**ADVANCED ORGANIC CHEMISTRY**
Blažej, A. & Košík, M.	**PHYTOMASS: A Raw Material for Chemistry and Biotechnology**
Brown, S.S. & Kodama, Y.	**TOXICOLOGY OF METALS: Clinical and Experimental Research**
Corey, E.R., Corey, J.Y. & Gasper, P.P.	**SILICON CHEMISTRY**
Dunn, A.D. & Rudorf, W.D.	**CARBON DISULPHIDE IN ORGANIC CHEMISTRY**
Foá, V., Emmet, E.A., Moroni, M. & Colonksi, A.	**OCCUPATIONAL AND ENVIRONMENTAL CHEMICAL HAZARDS: Cellular and Biochemical Indices of Monitoring Toxicity**
Fresenius, P.	**ORGANIC CHEMICAL NOMENCLATURE**
Hudlicky, M.	**CHEMISTRY OF ORGANIC FLUORINE COMPOUNDS: A Laboratory Manual, 2nd (Revised) Edition**
Hudlicky, M.	**REDUCTIONS IN ORGANIC CHEMISTRY**
Keese, R. & Muller, R.K.	**FUNDAMENTALS OF PREPARATIVE ORGANIC CHEMISTRY**
Kováč, J., Krutošíková, A. & Kada, R.	**CHEMISTRY OF HETEROCYCLIC COMPOUNDS**
Lukevics, E., Pudova, O. & Sturkovich, R.	**MOLECULAR STRUCTURE OF ORGANOSILICON COMPOUNDS**
Maksic, Z.B.	**MODELLING OF STRUCTURE AND PROPERTIES OF MOLECULES**
Manitto, P.	**BIOSYNTHESIS OF NATURAL PRODUCTS**
Mason, J.M.	**PRACTICAL SONOCHEMISTRY: User's Guide to Applications in Chemistry and Chemical Engineering**
Neidleman, S.L. & Geigert, J.	**BIOHALOGENATION: Principles, Basic Roles and Applications**
O'Hagan, D.	**THE POLYKETIDE METABOLITES**
Pizey, J.S.	**SYNTHETIC REAGENTS:**
	Volume 1: Dimethyl Formamide; Lithium Aluminium Hydride; Mercuric Oxide; Thionyl Chloride
	Volume 2: *N*-Bromosuccinide; Diazomethane; Manganese Dioxide; Raney Nickel
	Volume 3: Diborane; 2,3-Dichloro-5,6-Dicyanobenzoquinone DDQ; Iodine; Lead Tetra-acetate
	Volume 4: Mercuric Acetate; Periodic Acid and Periodates; Sulfuryl Chloride
	Volume 5: Thallium (III) Acetate and Trifluoroacetate; Ammonia; Iodine Monochloride
	Lithium Aluminium Hydride (reprinted from Synthetic Reagents, Volume 1)
Sakurai, H.	**ORGANOSILICON AND BIOORGANOSILICON CHEMISTRY: Structure, Bonding, Reactivity and Synthetic Application**
Smith, E.H.	**COMPARATIVE OXIDATION OF ORGANIC COMPOUNDS**
Smith, K.	**SOLID SUPPORTS AND CATALYSTS IN ORGANIC SYNTHESIS**
Sterba, V. & Panchartek, J.	**KINETIC METHODS IN ORGANIC CHEMISTRY**
Trinajstic, N., Knop, J.V., Müller, W.R. & Szymanski, K.	**COMPUTATION CHEMICAL GRAPH THEORY: Generation of Structures by Computer Methods**
Tundo, P.	**CONTINUOUS FLOW METHODS IN ORGANIC SYNTHESIS**
Vernin, G.	**THE CHEMISTRY OF HETEROCYCLIC FLAVOURING AND AROMA COMPOUNDS**
Lukevics, E. & Zablocka, A.	**NUCLEOSIDE SYNTHESIS: Organosilicon Methods**

NATURAL AND SYNTHETIC SWEET SUBSTANCES

ALŽBETA KRUTOŠÍKOVÁ, B.Sc., Ph.D.
and
MICHAL UHER, B.Sc., Ph.D.
Department of Organic Chemistry, Slovak Technical University
Bratislava, Czechoslovakia

Translation Editor: **Dr JOHN HUDEC,**
Ryvan Chemical Co. Ltd, Botley, Southampton, UK

ELLIS HORWOOD
NEW YORK LONDON TORONTO SYDNEY TOKYO SINGAPORE

English Edition first published in 1992
in coedition between
ELLIS HORWOOD LIMITED
Market Cross House, Cooper Street,
Chichester, West Sussex, PO19 1EB, England

A division of
Simon & Schuster International Group
A Paramount Communications Company

and
VEDA
Publishing House of the Slovak Academy of Sciences,
Klemensova 19, 814 30 Bratislava, Czechoslovakia

Distributed in East European countries, China, Cuba, Northern Korea, Mongolia, Vietnam by VEDA, Publishing House of the Slovak Academy of Sciences, Klemensova 19, 814 30 Bratislava, Czechoslovakia

Printed in Czechoslovakia

British Library Cataloguing in Publication Data

A catalogue record for this book is available from the British Library.

ISBN 0–13–612805–X
ISBN 80–224–0244–3 (Veda)

Library of Congress Cataloguing-in-Publication Data

is available from the Publishers.

In memory of our colleagues
Vierka Knoppová
and
Adolf Jurášek

Table of contents

Introduction

Although sweet tasting foodstuffs were always, in all probability, an integral part of human diet, it is the 'modern man' that suffers most from the consequences of high sucrose consumption; this is the result of technological advances in sugar production. The phenomena of over indulgence such as obesity, arteriosclerosis, an increasing number of diabetics and people with carious teeth are nowadays well known. Clearly, these are important enough reasons to motivate our search for more acceptable sweet substances that do not possess some of the above shortcomings of sucrose. As yet, no ideal sweetener, which is soluble, stable, non-carious, low in calories and non-toxic, has been found.

The problem of selecting a suitable sweetener and its optimal dosage involves many people — the consumers, health care workers and the food manufacturers. The issue is a complex one — nutritional, toxicological, technological as well as economic aspects must all be taken into consideration.

Numerous sweet tasting substances have been recorded in the literature, but only very few have been commercially successful. The remaining ones are either too toxic, or their sweet potency is too low.

The sweeteners that can be considered suitable for commercial exploitation can be crudely divided into two categories — the natural sweet substances and the synthetic (artificial) ones.

The first group encompasses mono- and oligosaccharides, starch hydrolysates and sugar alcohols, in addition to the non-saccharidic sweet substances such as proteins, terpenoids, steroidal sapogenins, dihydroisocoumarins and flavonoids.

It has been shown that plants contain factors that stimulate or suppress the sensation of sweet taste. Such factors are useful in the study of mechanisms that are responsible for eliciting the sweet taste on a molecular level. They can also be used as taste modifiers.

As the analytical instrumentation and synthetic methodology of organic chemists became increasingly more powerful, the structures of many natural sweet substances were elucidated, and many of them were in due course synthesized. As a consequence, many natural sources of sweet compounds have been abandoned and the border-line between the group of natural and synthetic sweet substances became less distinct (dihydrochalcone glycosides fall into this category).

The history of synthetic sweeteners, in comparison with the natural ones, is just over a hundred years old. The beginning has been traditionally set in 1878, the year when saccharin was discovered. Later, several other sweeteners were discovered and commercialized, for instance dulcin or cyclamates. However, the application of artificial sweeteners in food has always been rather controversial. Recent additions to the family of commercial synthetic sweeteners are oxathiazinone dioxides and the group of aspartame and its analogues.

The use of more than one sweetener at the same time has proved to be a fruitful idea. Not only is the net weight decreased in comparison with products that contain sucrose, it also often helps to modify and amplify the sweet taste (multiple or mixed sweeteners). The synergistic effect of mixed sweeteners is operative not only within the natural and synthetic groups but also in a combination of compounds from both groups.

The intention of this book is to address the general science reader, specialists in food technology, dieticians, research scientists and food analysts as well as teachers and to hope that they will find the subject of *sweetness* interesting.

We are indebted to all who made the publication of this book possible, above all to the translator Dr Peter Zálupský, to the publisher, Ellis Horwood, for his interest and support and to Dr John Hudec, the translation editor, for his generous help and expertise.

The authors

1

Structure–sweetness relationship

1.1 PHYSIOLOGY OF TASTE PERCEPTION

Taste has been generally acknowledged to be the principal attribute of quality in food. Unacceptable taste can disqualify the food on the market square even if it possesses high nutritive value or aesthetic appearance.

Taste perception is a complex, as yet not fully understood process. The taste sensation is elicited as a result of an interaction between the molecules of the taste-stimulating compound and the taste receptor. Taste receptors in humans are of several types that differ in their location on the tongue, and respond to four basic tastes — salty, sour, bitter and sweet (Fig. 1.1).

There is a belief in Asia that in addition to sweetness, saltiness,

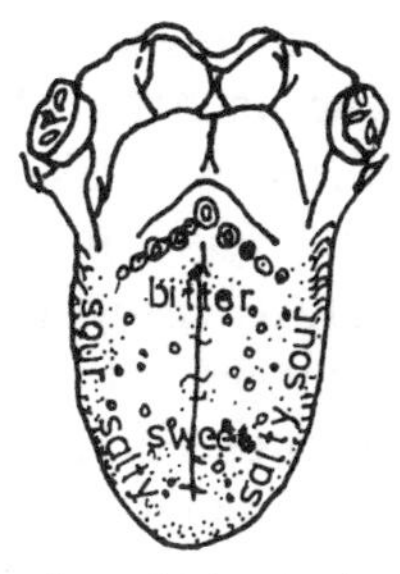

Fig. 1.1 — Location of taste buds on the tongue.

sourness and bitterness there is a fifth fundamental taste, called *umami*. It is related to the monosodium salt of glutamate (MSG) and its related 5′-nucleotides. This type of taste has presently reached the best level of understanding and definition [1, 2].

Sweet taste receptors reside in porous taste buds that are located at the tip of the tongue [3]. The cups have a uniform structure which consists of supporting cells and 10–15 hairy receptor cells per cup [4, 5], which are responsible for sensing the taste. However, the information on the function and structure of taste cups is still rather fragmentary.

Protein molecules have been recognized to form the structures that are responsible for the interaction with the stimulant molecule. They have been isolated from the epitels of the bovine tongues and their complexes with sugars studied [6].

Similar complexes have been found to be formed when protein fractions isolated from tongues of other mammals (apes, rats, cows, cats) were mixed with a sweet taste stimulant [7–13].

The first attempts at the generalization of the structure–sweetness relationship stressed the physico-chemical nature of the interaction between the stimulant and the receptor. Thus Beidler [12–14] interpreted the taste as the consequence of a selective adsorption on the taste receptor and used the adsorption isotherm for its description

$$\frac{C}{R} = C\frac{1}{R_{\mathrm{m}}} + \frac{1}{KR_{\mathrm{m}}},$$

where R is the response at the concentration C of the stimulant, R_{m} the maximal response on fully saturated receptors, and K the constant that is characteristic for the interaction of a particular stimulant and the receptor.

The local geometry of the surface membrane and its immediate environment changes as a result of the adsorption. This initiates a cascade of events which lead ultimately to the sensation of taste. Beidler's hypothesis was substantiated later [15].

Although the theory indicated the most probable mechanism of interaction, it was still vague on details [16]. One such crucial point hinged on the question of the structural variation at the receptor site. In the face of a large number of types of sweet substances with differing chemical structures, it was most reasonable to assume that there must be also a number of differently shaped receptors. A large body of data was accumulated during the studies on the structural elucidation of the receptor, all confirming the assumed structural variation [17–19]. The receptors were found to be species specific, thus baboons and rats differed in their response to the identical stimulant, dihydrochalcone-neohesperidin [20]. Corroborative evidence has been obtained from experiments on local stimulation of human tongue by various types of sweet substances [21]. Disparate responses to amino acids have been detected in humans and in rats [22]. Another experimental result dealt with the effect of two inhibitors on the taste response of various sweeteners [23]. Alloxan was shown to be able to suppress the sweet taste of sugars, but was neutral with synthetic sweeteners [24]. Numerous psychophysical studies confirmed the existence of several receptor sites and also the fact that their number in each species is unique [25–27].

The taste stimulus is believed to be transmitted from the receptor site along the nerve paths to the taste centre in the brain, which is located in the sensorial part of the cortex, in the vicinity of the so-called postcentral brain fold.

1.2 RELATIONSHIP BETWEEN CHEMICAL STRUCTURE AND SWEETNESS

The prime reason for our extensive background knowledge about the sweet taste is due to the increasingly important role that sweet substances have assumed in the human diet and in particular, with the harmful effects of excessive consumption of sugar on health.

Saccharides have been considered for a long time as the embodiment of sweet taste, and sweetness their most characteristic attribute. However, the advances in the chemistry of saccharides have revealed that they are far from being universally sweet; indeed their taste occupies almost the whole range from sweetness to bitterness.

The similarities in chemical structure such as between saccharides and acyclic polyalcohols, for instance, ethylene glycol and glycerine which vary only in the number of hydroxyl groups, suggested themselves quite naturally as the cause of the sweet taste in both groups of compounds [28]. However, the number of hydroxyl groups alone was obviously erroneous when sugars possessing the same number of hydroxyl groups, such as glucose or galactose, both saccharides with the same number of hydroxyl groups, demonstrated widely varying sweetness. Moreover, pentaols like xylitol, were found to be sweeter than polyols with six hydroxyl groups represented, for instance, by sorbitol [28].

The concept of sweetness based on the number of hydroxyl groups alone, clearly necessitated a modification which would take into account other factors as well. This was the basis of the theory of sweetness put forward by Oertly and Myers [29] who utilized the concept of *sapophoric* groups, first introduced by Cohn [30]. They explained sweetness as being due to the presence of a pair of sapid functions, designated *auxoglucs* and *glucophores*, by analogy to auxochromes and chromophores of the coloured species. The authors [29] believed that the sweet taste was manifested only if both taste functions, represented by the respective chemical (partial) structures were present. However, they failed to show how the synthetic sweet substances fitted into the scheme, and they did not propose any mechanism, however tentative, for the interaction of 'their' sweet substances with the receptor.

Other theories appeared in due course which brought into consideration other factors which were suspected of being related to sweetness, such as 'vibratory' hydrogen [31], resonance energy [32], hydrophobicity and Hammett constants [33].

Only when the ring structure of saccharide molecules was established in the 1930's [34], it became clear that the differences in sweetness of diastereoisomers could often be attributed to the configurational change at a single carbon atom. This has been demonstrated for a number of sugars, for instance α-D-glucopyranose was found to be sweeter than α-D-galactopyranose, just as β-D-mannopyranose was bitter whilst α-D-mannopyranose was sweet (Scheme 1.1).

In the absence of any unifying theory of sweetness, the controversial issue was admirably assessed by Verkade [35] who reviewed the sweet-

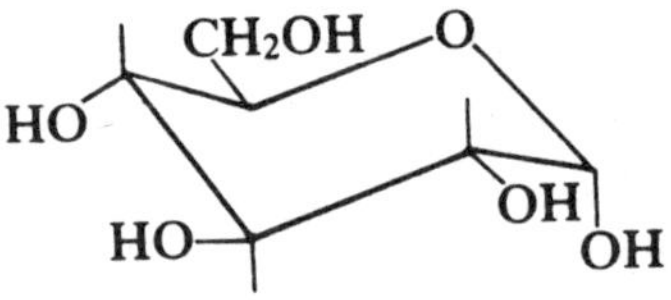

α-D-glucopyranose
(sweet)

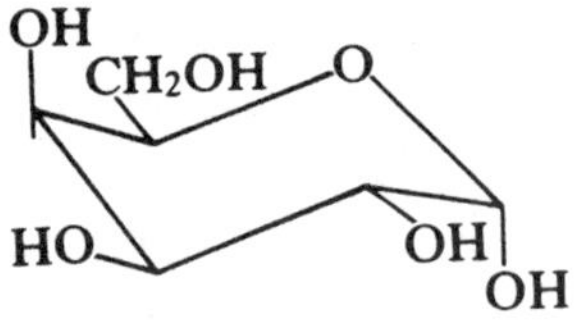

α-D-galactopyranose
(less sweet)

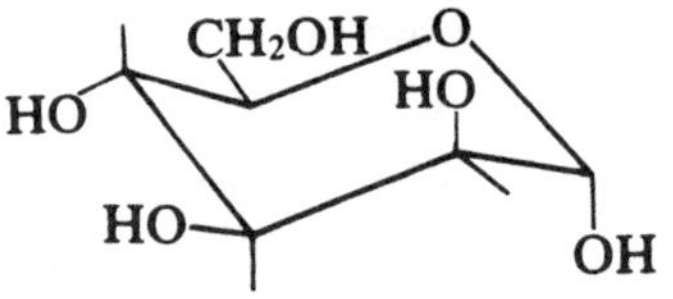

α-D-mannopyranose
(sweet)

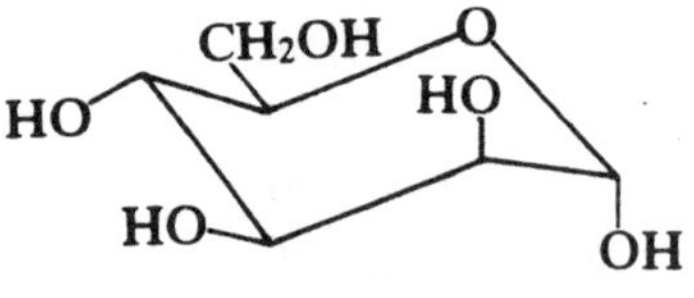

β-D-mannopyranose
(bitter)

Scheme 1.1

ness and bitterness of several analogous organic compounds. Even though the hydrogen bond forming capability of sugar molecules is omnipresent and forms the basis of all their typical properties, a hypothesis of sweetness based on hydrogen bonding did not appear until the 1960's.

1.2.1 Shallenberger's hypothesis

The enormous variation in the structure of compounds with sweet taste impeded for a long time the recognition of the fundamental structural fragments that are capable of interacting with the receptor and thus eliciting the sensation of sweet taste. It was Shallenberger, an expert in sugar chemistry [36–39], who was the first to attempt to define the probable mode of interaction at the receptor site [40]. He postulated the interaction as being due to hydrogen bonding between the stimulant and the active centres on the receptor. He suggested further that the interaction was highly stereospecific and was the first step in a series of dynamic chemical events that eventually resulted in the perception of sweetness.

Although Shallenberger originally tailored his hypothesis only for sugars, he extended it successfully later to other groups of sweet substances which were chemically and structurally quite different from sugars [39–41]. The most valuable idea in Shallenberger's hypothesis was the concept of the 'active' part of the sweet molecule as the AH–B system. The A and B represented electronegative atoms and AH an acidic function. Provided A and B are correctly oriented and at the correct separation, B will act as a base and a hydrogen bonding will ensue. The AH–B system (Fig. 1.2) of a sweet molecule can interact

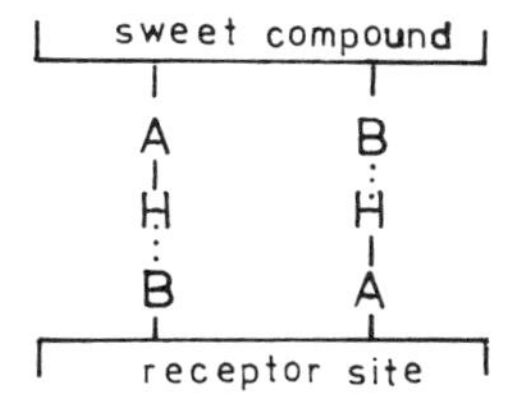

Fig. 1.2 — Intermolecular hydrogen bond between the sweet compound and the receptor [41].

with another AH–B system, such as one present in a protein molecule at the receptor site (taste bud), resulting in a loose complex. The two new intermolecular hydrogen bonds activate both the onset and the intensity of the sweet stimulus. The AH–B system in sugars is represented by the intramolecularly hydrogen-bonded *α*-glycol fragment. Shallenberger and Acree [39, 40] used this concept to analyse various types of sweet compounds in order to establish whether they contained the intramolecular AH–B system (Scheme 1.2).

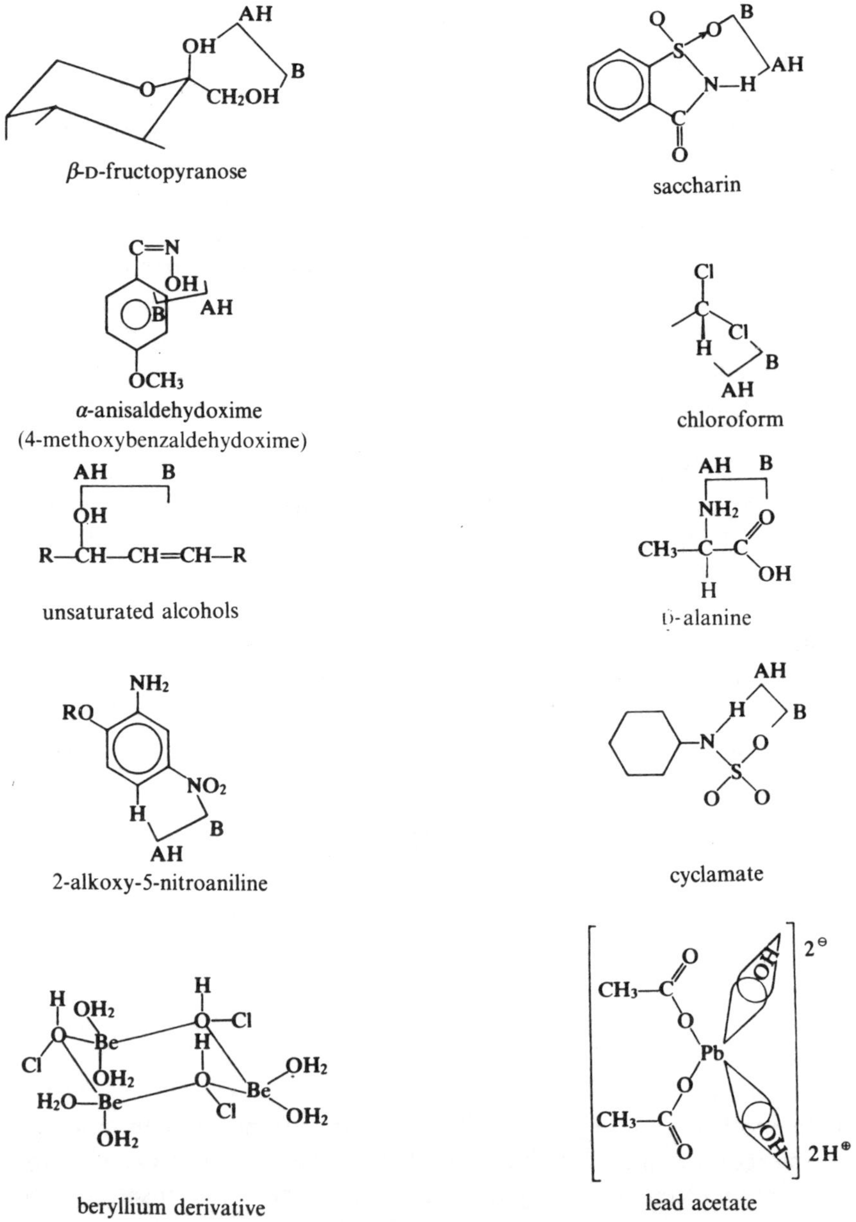
AH
OH
B
O
CH₂OH
β-D-fructopyranose
O
S
O
B
AH
N—H
C
O
saccharin
C=N
OH
B
AH
OCH₃
α-anisaldehydoxime
(4-methoxybenzaldehydoxime)
Cl
C
Cl
H
B
AH
chloroform
AH B
OH
R—CH—CH=CH—R
unsaturated alcohols
AH B
NH₂
O
CH₃—C—C
OH
H
D-alanine
NH₂
RO
NO₂
H
B
AH
2-alkoxy-5-nitroaniline
AH
H
B
N
O
S
O O
cyclamate
H OH₂
O
Be
Cl
OH₂
H₂O—Be
OH₂
H
O—Cl
H
O
Be
Cl
OH₂
OH₂
beryllium derivative
O
CH₃—C
O
Pb
O
CH₃—C
O
OH
OH
2⊖
2H⊕
lead acetate

Scheme 1.2

β-D-Fructopyranose, one of the compounds analysed, has the most acidic hydrogen atom on the anomeric OH group, representing the AH function, and the adjacent rotatable hydroxymethyl group which serves as the B component in the intramolecular AH–B system.

The application of the above principles to saccharin ascribes the AH role to the NH group; there are two choices for B, either the carbonyl group oxygen atom or the oxygens of the sulphonyl group. The latter alternative is most likely because the enol form, pseudosaccharin, lacks sweetness.

Chloroform may serve as an example of a small sweet molecule. The electron withdrawing effect of the trichloromethyl group raises the acidity of the hydrogen atom and thus acts as the AH unit, with the chlorine atom of a molecule as the B counterpart of the AH–B system.

Similar arrangement is repeated in unsaturated allylic alcohols. This makes the hydrogen atom of the hydroxyl group weakly acidic and thus an AH unit whilst the double bond acts as a base (Scheme 1.2).

2-Alkoxy-5-nitroaniline illustrates well the group of compounds known as sweet benzene derivatives. The sweetness in this series of nitrobenzenes depends to a large extent on the character and the position of other substituents.

Shallenberger concluded that the geometry of the AH–B system, both in saccharides and on the active site, governed the stability of the resultant complex, thus making the configuration and conformation of sweet molecules two crucial factors for eliciting the sweet response. Rigid molecular structures that contain aromatic residues, such as nitroanilines and saccharin, have a decided advantage because their interorbital A–B separation is fixed at a correct distance.

It is known that artificial sweeteners are often several hundred times sweeter than sugars. Furthermore, the sweet taste of sugars persists only for seconds as the stability of the interaction complex is low. In order to enhance it, and achieve a two-point interaction mediated by hydrogen bonds, the distances between A and B must correspond to the distances of their partner groups at the receptor and be of the order of ~ 0.3 nm.

Scheme 1.2 depicts one plausible interpretation of the sweetness of inorganic compounds; however, no geometrical data are available for these substances. An important role is clearly played by water molecules that are illustrated in both cases.

Birch and Lee [42] assert that if sweet substances and active sites at the receptor possess matching geometries, it will be the rate of formation of the complex, rather than its stability (lifetime), that will control the intensity and the duration of the sweet sensation.

1.2.2 Extension of Shallenberger's hypothesis

Although far from perfect, Shallenberger's hypothesis succeeded to rationalize the mode of interaction of many types of sweet substances with the receptor. Nevertheless, several independent studies [43–45] appeared to suggest the necessity for a third binding site in the molecule of a sweet substance as a prerequisite of a potent sweet taste response. The fact that many D-amino acids are sweet, whereas their L-counterparts were tasteless or even bitter, indicates the high stereospecificity of the receptor (Tables 1.1, 1.2) [46, 47].

Table 1.1 — Taste of amino acids [47]

No.	Compound	Taste	Threshold value[a]
1	D-Alanine	sweet	12–18
2	L-Alanine	sweet	12–18
3	2-Amino-4-pentenoic acid	sweet	8–12
4	D-2-Aminobutanoic acid	sweet	12–16
5	L-2-Aminobutanoic acid	bitter-sweet	12–16
6	2-Aminoisobutanoic acid	sweet	5–10
7	D-Arginine	neutral	–
8	L-Arginine	bitter	–
9	D-Asparagine	sweet	3–6
10	L-Asparagine	neutral	–
11	D-Aspartic acid	neutral-to-sweet	–
12	L-Aspartic acid	sour-to-neutral	–
13	L-Azetidinecarboxylic acid	sweet	6–8
14	Betaine	sweet	30–50
15	Cycloleucine	bitter-sweet	3–6
16	D-Cysteine	sulphurous	–
17	L-Cysteine	sulphurous	–
18	D-Cystine	neutral	–
19	L-Cystine	neutral	–
20	D-3,4-Dihydroxyphenylalanine	sweet	1–2
21	L-3,4-Dihydroxyphenylalanine	bitter	–
22	D-Glutamine	sweet	8–12
23	L-Glutamine	neutral	–
24	D-Glutamic acid	sour-to-neutral	–
25	L-Glutamic acid	sour	–
26	Glycine	sweet	25–35
27	D-Histidine	sweet	2–4
28	L-Histidine	bitter	–
29	L-Homoserine	neutral	–
30	D-allo-4-Hydroxyproline	neutral	–
31	L-4-Hydroxyproline	sweet	–
32	L-allo-4-Hydroxyproline	neutral	5–7
33	D-5-Hydroxytryptophan	sweet	0.2–0.6
34	L-5-Hydroxytryptophan	bitter	–

Table 1.1 (continued)

No.	Compound	Taste	Threshold value[a]
35	D-Isoleucine	sweet	8–12
36	L-L-Isoleucine	bitter	–
37	D-Leucine	sweet	2–5
38	L-Leucine	bitter	–
39	D-Lysine	sweet	–
40	L-Lysine	bitter-sweet	–
41	D-Methionine	sulphurous sweet	4–7
42	L-Methionine	sulphurous	–
43	D-Norleucine	sweet	5–8
44	L-Norleucine	bitter	–
45	D-Norvaline	sweet	3–5
46	L-Norvaline	bitter	–
47	D-Ornithine	sweet	–
48	L-Ornithine	sweet	–
49	D-Phenylalanine	sweet	1–3
50	L-Phenylalanine	bitter	–
51	D-Phenylglycine	sweet	10–14
52	L-Phenylglycine	bitter	–
53	D-Pipecolinic acid	neutral	–
54	L-Pipecolinic acid	bitter	–
55	D-Proline	neutral	–
56	L-Proline	bitter-sweet	25–60
57	Sarkosine	sweet	15–20
58	D-Serine	sweet	30–40
59	L-Serine	sweet	25–35
60	D-Threonine	sweet	40–50
61	*l*-Threonine	sweet	35–45
62	D-Tryptophan	sweet	0.2–0.4
63	L-Tryptophan	bitter	–
64	D-Tyrosine	sweet	1–3
65	L-Tyrosine	bitter	–
66	D-Valine	sweet	10–14
67	L-Valine	bitter	–

[a] In μM/ml, threshold concentration of sucrose is 10–12 M/ml [47]. Perception threshold — the lowest concentration of a compound, detectable by human sensors.

Kier [48] finally identified and characterized the third binding site (X) and thus extended the original AH–B sweet pharmacophore concept by the third lipophilic function.

The validity of the three-point interaction was appreciated by Shallenberger [49] also; he accordingly modified his earlier theories and utilized it for rationalizing the difference in the sweetness of enantiomers. However, whilst Kier imagined the third binding site as being

Table 1.2 — Taste of amino acids derivatives and other related compounds [47]

No.	Compound	Taste	Threshold value
1	Acetamide	neutral	–
2	*N*-Acetylglycine	sour-to-neutral	–
3	*N*-Acetyl-D-phenylalanine	sour-to-neutral	–
4	2-Aminoethanol	neutral	–
5	D,L-3-Aminobutanoic acid	sweet	100–300
6	4-Aminobutanoic acid	neutral	–
7	6-Aminocaproic acid	bitter	–
8	3-Aminopropionic acid	sweet	1000–1400
9	Benzamide	bitter	–
10	*N*-Benzoylglycine	sour-to-bitter	–
11	Glycine methyl ester	neutral	–
12	Glycyl-glycine	neutral	–
13	D-Phenylalanylamide	bitter	–
14	D-Phenylalanine methyl ester	bitter	–
15	L-Phenylalanylglycine	bitter	–
16	*N*-Phenylglycine	neutral	–
17	Pyrrolidine	bitter	–

the centre of a dispersed bond designated X, Shallenberger and Lindley [50] and Shallenberger [51] described it as a lipophilic centre combined with a hydrophobic function, and designated it γ (Fig. 1.3a, b).

As can be seen from Fig. 1.3, the value of γ is indicated by point 1 for sugars and amino acids (Fig. 1.3a), whilst point 2 pertains to dipeptide sweeteners (Fig. 1.3b).

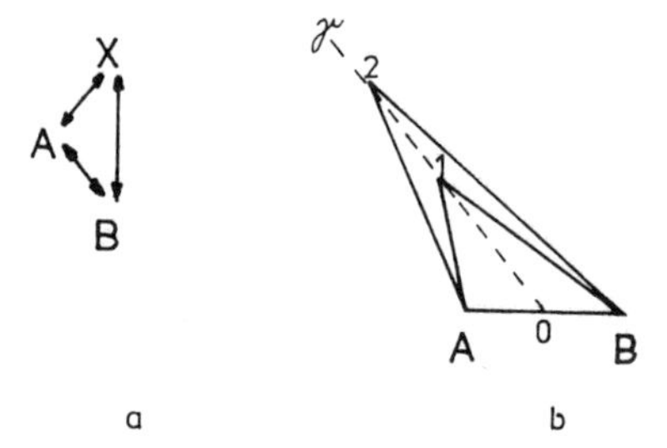

Fig. 1.3 — (a) A three-point sweet unit (AH–B–X). AB = 2.6 nm, AX = 0.35 nm, BX = 0.55 nm [48]. (b) Geometrical relations between the AH–B and the third component γ [51].

The sweetness of molecules has been examined not only from the viewpoint of their geometry, many authors attempted to correlate it also with the physical properties. Thus Deutsch and Hansch [43] and Ferguson and his group [52–56] successfully correlated the relative sweetness of 2-substituted 5-nitroanilines with the Hammett constants

$$\log RS = -k\pi^2 + k^1\pi + k^2\sigma + k^3, \tag{1}$$

where RS represents relative sweetness, σ the Hammett constant; constant π characterizes the free energy of the substituent and was defined as hydrophobic bonding constant $\pi = \log P_X/P_H$ [57]. The P values represent the distribution coefficients between ethanol and water of unsubstituted P_H and P_X-substituted compounds. The sweetness values of some 5-nitroanilines relative to sucrose, as determined by Blanksma and Hoegen [58] are presented in Table 1.3.

Table 1.3 — Relative sweetness of 2-substituted-5-nitroanilines

X	σ	π	log RS observed	log RS calculated	$\Delta \log RS$
H	0	0	1.602	1.729	−0.127
CH_3O–	−0.27	0.02	2.519	2.193	0.327
C_2H_5O–	−0.24	0.48	3.146	2.942	0.205
n-C_3H_7O–	−0.24	0.98	3.699	3.746	−0.047
F	0.06	0.14	1.602	1.845	−0.243
Cl	0.23	0.71	2.602	2.451	0.151
Br	0.23	0.86	2.903	2.693	0.120
I	0.28	1.20	3.097	3.149	−0.052
CH_3	−0.17	0.56	2.519	2.942	−0.424

log RS was calculated from equation (2) [58].

Blanksma and Hoegen also derived an empirical equation which relates relative sweetness and Hammett constant

$$\log RS = 1.610\,\pi - 1.832\,\sigma + 1.729, \tag{2}$$

$n = 9$, $r = 0.936$, $s = 0.383$.

The π-values and relative sweetness are directly proportional as can be seen from data in Table 1.3. Methoxy-, ethoxy- and propoxy-substituted compounds aptly exemplify this relationship. Deutsch and Hansch [33] stressed that the ability to form hydrogen bonds and solubility in water are necessary prerequisites in order that compounds manifest their sweetness. However, this should not detract from the importance of the hydrophobic part that is responsible for the interaction with the taste receptors. The interaction itself is additionally influenced by several factors such as polarity, distance between charges within the molecule, electron density, the Hammett constant σ, as well as by steric factors. Redistribution of charges in nitroanilines becomes a factor either when there is an electron donating substituent in *para* position to the nitro group or when there is an electron withdrawing substituent in *meta* position with respect to the amino group (Table 1.3).

Extensive studies of nitroanilines led Kier to propose a hypothetical mechanism of interaction of these compounds with the receptor. He assumed the presence of a tryptophane residue at the receptor site in such an orientation that the hydrophobic interactions with the substituent of 2-nitroanilines would be maximized.

Provided that the three-point interaction between the sweet stimulant and the receptor is strong enough, conformational changes in the proteins can be induced, which in turn sets off a chain of events that result ultimately in sweet sensation.

Steric parameters of sweet molecules, specially the interatomic distances in the AH–B unit, have been determined by a computer program STERIMOL [64]. The software calculates the length of the molecule, five width parameters, plots *xy* and *yz* plane projections of the molecule and gives the spatial coordinates of atoms as well. When used for the analysis of 10 nitroanilines and 6 cyanoanilines, the program indicated a significant relationship between steric factors and activity for 5 nitroanilines and 5 cyanoanilines.

Ferguson developed an argument based on his earlier studies on sweetness of trisubstituted benzene derivatives [52, 56], compounds of the type

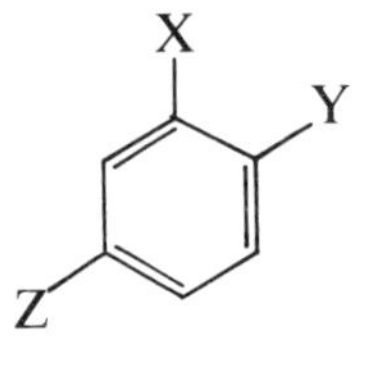

X
Y
Z

X = NH_2, OH, CH_3
Y = NH_2, H, halogen (*ortho*- and *para*-orientating substituents)
Z = NO_2, CN, CF_3

interact with the receptor in such a way that X and Y act as the classical AH–B moiety and the Z-phenyl takes part in a hydrophobic interaction with the receptor. However, practical experiments demonstrated that sweet compounds of the above type often contained either a nitro, carboxy or a cyano group — the interchange of these groups influenced substantially the sweet potency. Thus it appears as if the receptor was equipped with respective structures into which these groups fit [67, 68]. However, this contradicts the conclusion of Shallenberger and Acree [44] who suggest that both the NO_2 group in 2-propoxy-5-nitroaniline and the $COO^{\ominus}$ group in amino acids (tryptophane) interact with the same receptor site.

Detailed study of the mechanism of action of the intensely sweet urea derivate Suosan and related compounds [67–69] revealed that these sweeteners interact with the taste receptor by three structural elements: the hydrogen bond originating at the acidic AH group

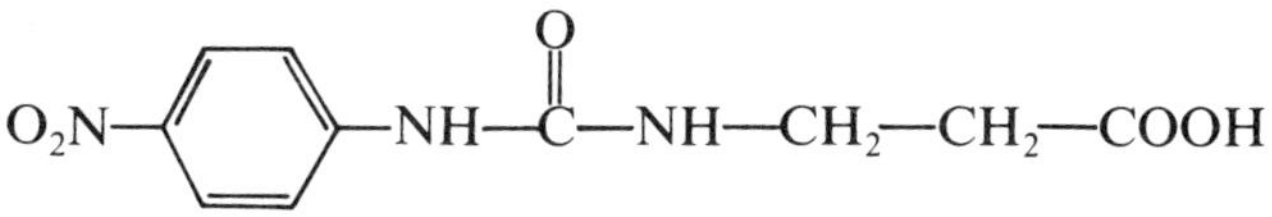

$$O_2N-C_6H_4-NH-\overset{O}{\overset{\|}{C}}-NH-CH_2-CH_2-COOH$$

N-(4-nitrophenyl)-*N′*-(2-carboxyethyl)urea (Suosan)

(NH group of Suosan), B group that carries a partial negative charge that is responsible for the ionic interaction ($COO^{\ominus}$) and finally by the D group NO_2 or CN that act as the acceptor partner for the hydrogen bond.

In order to achieve maximum interaction with the receptor, $NH^{\oplus}$ and $CH_2CH_2COO^{\ominus}$ groups have to assume a *cis*-configuration, other interaction groups must have fixed an optimal mutual separation (Fig. 1.4).

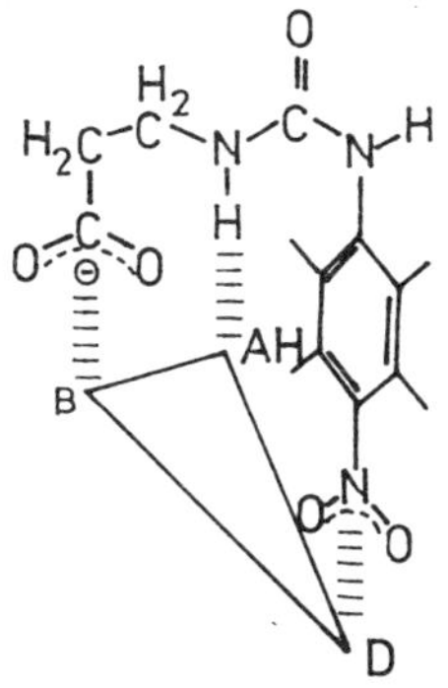

Fig. 1.4 — Suosan — centres of the three-point interaction with the receptor [69]. AB = 0.25 nm, AD = 0.55 nm, BD = 0.60 nm.

1.2.3 Structure–sweetness relationship in dipeptides

The relationship between the sweet potency of dipeptides and their degree of hydrophobic properties in relation to the geometry of their side-chains have been thoroughly studied [70].

If we try to apply the Shallenberger theory to dipeptides, we can immediately recognize ammonium and carboxy groups, $NH_4^{\oplus}$ and $COO^{\ominus}$, of the aspartyl residue as the AB–H system. Apparently, these groups are the first ones in contact with the receptor site. This interaction results in the required conformational changes in order that additional non-bonding interactions can take place.

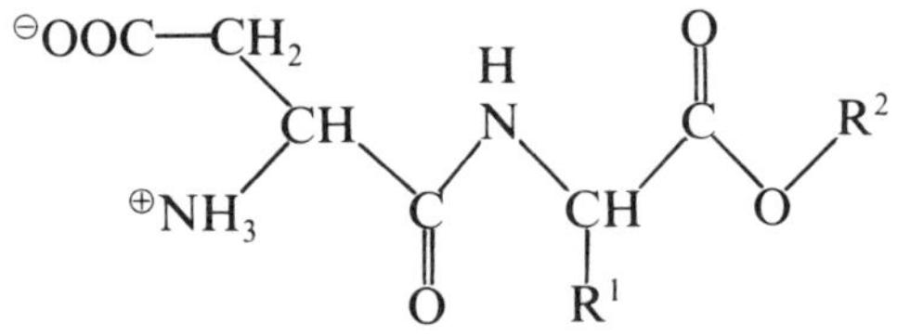

$^{\ominus}OOC{-}CH_2$ H O
CH N C R^2
$^{\oplus}NH_3$ C CH O
O R^1

Hydrophobic groups R^1 and R^2 in the above formula affect the interaction by their bulk. However, if they are too bulky, they can create a steric barrier, thus preventing the AB–H system from coming into contact with the receptor [79].

Lelj and his group undertook a number of experimental and theoretical studies aimed at the clarification of the shape of the 'active' three-dimensional aspartame structures. Proton NMR spectra enabled them to assess the dihedral angles in the side-chains of L-Asp-L-Phe-OMe, whilst semiempirical quantum mechanical calculations were used to 'predict' the likely conformations in aqueous solution. Lelj and co-workers assumed that the most abundant conformations were the ones that fitted best into the receptor. As it turned out, such a conformation was characterized by a planar arrangement, the plane also contained the $NH_4^{\oplus}$ and $COO^{\ominus}$ groups. It is interesting to note that the part of the molecule that lies immediately above the polar groups is remarkably flat. The best possible 'lock' for such a flat 'key' would be a narrow groove at the receptor site into which the molecule of the sweet substance could be inserted as far as the polar groups would allow. Steric requirements of a flat molecule would be satisfied best by derivatives similar to L-aspartic acid or aminomalonic acid. Clearly, the derivatives of L-glutamic acid do not fit and are therefore inactive.

Conformational analysis of aspartame showed that the group of 9 possible conformers included the acidic, basic and zwitterionic forms of aspartame. It follows that, for any ionic forms, the proportion of any one of the conformers does not exceed 30 %, thus suggesting high conformational flexibility of aspartame in a broad range of pH values. Three of the nine possible conformers were judged to be able to participate in binding to the receptor *via* a suitable AH–B system. However, two of these had to be eliminated on steric grounds. The remaining conformation that is responsible for eliciting sweet taste is shown already located at the tentative active centre of the receptor (Fig. 1.5).

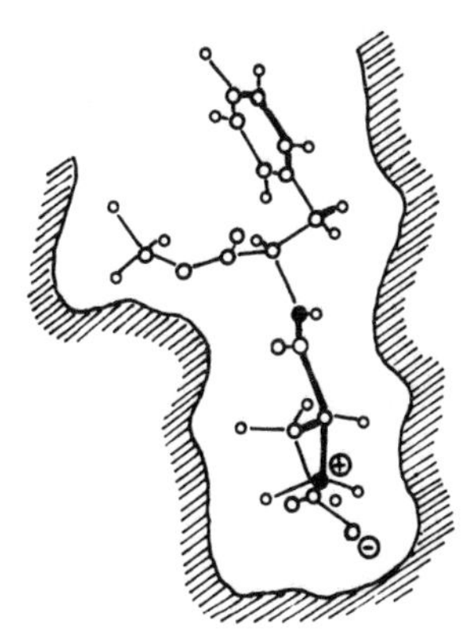

Fig. 1.5 — Simplified picture of an interaction of aspartame with the receptor [72, 73].

1.2.4 QSAR analysis of sweeteners

QSAR has proved to be a valuable tool in probing the sweetness of molecules. Its scope is based on the observation that electrostatic intramolecular energy ΔE of a pseudohydrogen bond H–B correlates with the sweet potency (*SP*) [75]. The plot of *SP* against ΔE is a parabola with an optimum when E is in the range of 7.9–8.4 kJ/mol. This conclusion led to speculations that ΔE might be inversely proportional to the strength of intermolecular hydrogen bonds between the AH–B system and the receptor. Moreover, a great number of AH–B systems might cause steric inversion in order to form a strong hydrogen bond between A–H and the oxygen of the (peptide) carbonyl group and between B and (peptide) NH group, both localized in the middle of the β-turn conformation [76].

Iwamura [65] applied the QSAR analysis to a set of 49 perillartine derivatives. The activity of perillartines (log A) was expressed as the logarithm of the taste potency, disregarding the quality of the taste [77]. The spatial dimensions of molecules were assessed by the application of STERIMOL parameters [64]. The above approach cannot be considered a rigorous three-dimensional analysis.

OH
H N
C
R^1

Molecules were characterized by several parameters (Fig. 1.6). Thus L designates the length of the substituent R^1 measured along the bond that connects R^1 to the aldoxime C atom of perillartines; W_e, W_r, W_u and W_d, shown in Fig. 1.6, are the widths of molecules measured perpendicularly to the L axis [65]. The W parameters correspond to the original Verlop's B_n values (n = 1, 2, 3, 4) [64].

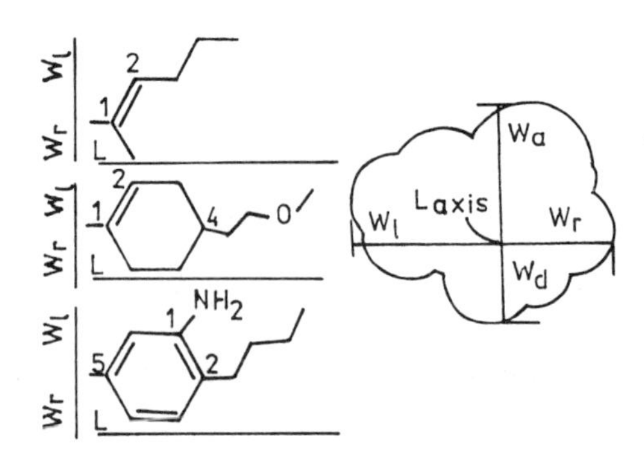

Fig. 1.6 — Space parameters L, W_l, W_r, W_a and W_d [65].

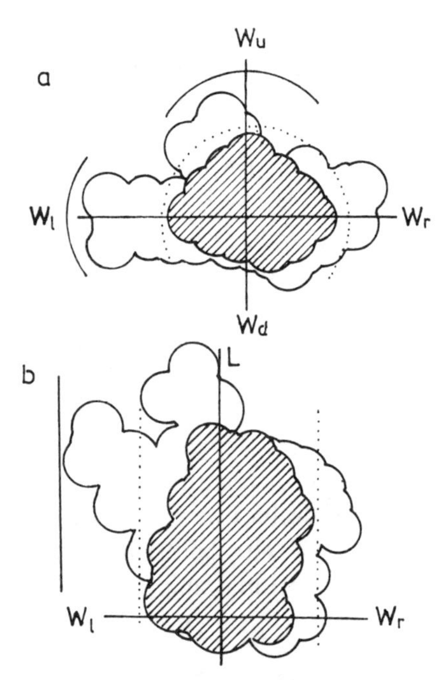

Fig. 1.7 — Perillartine — receptor bonds model. Maximal contours of the sweet perillartine derivative are striped. Unstriped part is one of the bitter analogues. Solid lines represent boundaries of the taste-eliciting part of the molecule, broken lines depict bitter barriers. (a) View from the C_1 towards the L axis. (b) Side view [65].

The empirical equation that fits best the experimental data for perillartines (incorporating all the above parameters) is

$$\log A = 0.63 \log P(\pm 0.31) + 0.19\,L(\pm 0.10) - 0.48\,W_e(\pm 0.18) - 0.62\,W_u(\pm 0.24) + 2.87(\pm 0.78)$$

$n = 38, r = 0.91, s = 0.32$ (n — number of compounds, r — correlation coefficient, s — standard deviation).

The assessment of hydrophobicity as $\log P$, where P is the distribution coefficient in 1-octanol–water mixture, was carried out by the method described by Hansch and Leo [78].

The results indicate that the ability of a molecule to elicit a taste sensation increases with an increase in $\log P$ and thus also with the increase in the molecular size. However, as the cross-section of the molecule W_u increases, the activity decreases. It follows that perillartines must be lipophilic if they are to be sweet, and should have a long, broad and flat molecular structure. Figure 1.7 illustrates such molecules in an interaction with the active site of the receptor [65].

Utilizing the methodology developed in the studies of perillartines, Iwamura [79] tested it on four series of α-aspartyl dipeptides.

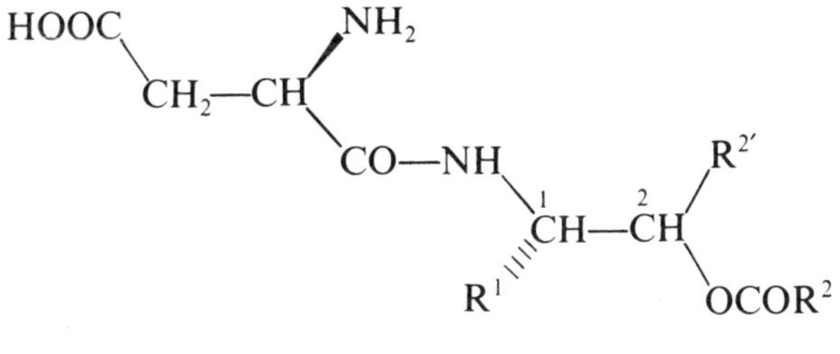

HOOC—CH_2—CH(NH_2)—CO—NH—$\overset{1}{C}$H(R^1)—$\overset{2}{C}$H($R^{2'}$)—$OCOR^2$

L-aspartylaminoethyl esters

A multidimensional linear regression equation was set up for L-aspartylaminoethyl esters

$$\log SP = 0.67\sigma^*(\pm 0.48) + 3.36L_2(\pm 1.03) - 0.29L_2^2(\pm 0.08) + 4.18(W_u)_1(\pm 0.88) - 0.85(W_u)_1^2(\pm 0.18) - 0.53L_1(\pm 0.18) - 11.33,$$

$n = 51,\ r = 0.88,\ s = 0.27.$

The influence of the steric effect of the substituent R^1 on the sweet taste has been reduced to the parameters of length (L_1) and width $(W_u)_1$, the length parameter (L_2) relates to the ester group $OCOR^2$. The electronic parameter σ^* characterizes the peptide bond and its values have been abstracted from the literature or estimated [78].

Two factors emerged as being essential for the manifestation of sweet taste from Iwamura's analysis of the four groups of L-aspartyl dipeptides (L-aspartylaminopropionates, L-aspartyl acid amides, L-aspartylaminoethyl esters, and L-aspartylaminoacetates). The first one was the electron withdrawing effect of substituents in the vicinity of the peptide bond and the overall spatial dimensions of the molecule as the second. Considering the σ^* regression coefficients, Iwamura [79] concluded, on the basis of QSAR analysis, that L-aspartylaminoacetates satisfy most satisfactorily the topology of the receptor's active site.

Van der Heijden and co-workers [80] attempted to characterize the stereochemistry of the interaction between a sweet substance and a receptor by a novel concept that utilized α, δ and ω parameters and the S-value. The parameters do not represent any steric features of the sweet molecule, they designate respectively the minimum, optimum and maximum distances for a three-point contact between points (preferably small surfaces) on the receptor site and the AH–B unit in the homologous series of sweeteners (nitroanilines, sulphamates, oximes, isocoumarins and dipeptides). The molecules in contact with the receptor are then characterized by the S-value, which represents the minimum distance between a particular atom (in α, δ or ω) and the plane defined by atoms A, H and B of the AH–B unit (Fig. 1.8).

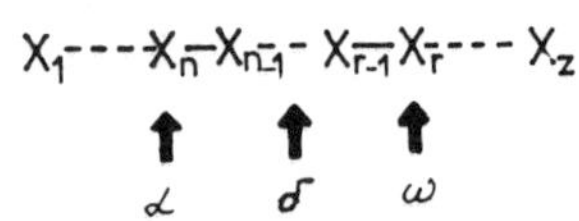

Fig. 1.8 — New concept parameters α, δ and ω, representing minimal, optimal and maximal distance between the hydrophobic third centre and hydrophilic AH–B moiety in sweeteners. X_1, ..., X_z represent the side chain of the members of homologous series of sweeteners.

Summarizing, we can say that all models [16, 80] point to the existence of several types of sweetness receptor.

1.2.5 Structure–sweetness relationship in saccharides

In the absence of any simpler model for the rationalization of sweetness than Shallenberger's hypothesis, many researchers have used the AH–B concept in their attempts to account for the structural subtleties of closely related molecules. The simple way to confirm Shallenberger's hypothesis is to modify the structure and taste of the product; this approach was at first limited to non-toxic saccharides. An inversion of configuration at one or more chiral centres of a saccharide molecule is liable to generate instability which can result in a change of conformation. Consequently, the chemical modifications were thus limited to the systems containing the glucopyranose moiety.

The non-reducing oligosaccharides are possibly the best models. On the other hand, glycosides have proved in many cases unsuitable because of the contamination by aglycones, which often confer an additional bitter taste [42].

The ideal sugar molecule for taste studies appears to be *α,α*-trehalose (mushroom sugar). This non-reducing disaccharide contains two glucopyranose units in their most stable $^4C_1$ conformation (all hydroxyl groups in equatorial orientation), linked glycosidically through their anomeric carbon atoms. Early work on trehalose [81] and related sugars contended that of all chemical modifications of the molecule, only substitution in position 4 appeared to reduce sweetness.

Considering the statistical odds, the most logical approach to finding the model AH–B systems would be the systematic selective removal of oxygen atoms, one by one, on the periphery of the sugar molecule, and tasting the product at each step. In fact, when this was done with the *α,α*-glucopyranoside, their sensory properties showed a remarkable similarity in each transformation. In contrast to early work on trehalose [81], later authors [82] found that removing the oxygen atom from position 3 decreased the sweetness and some monodeoxy deriva-

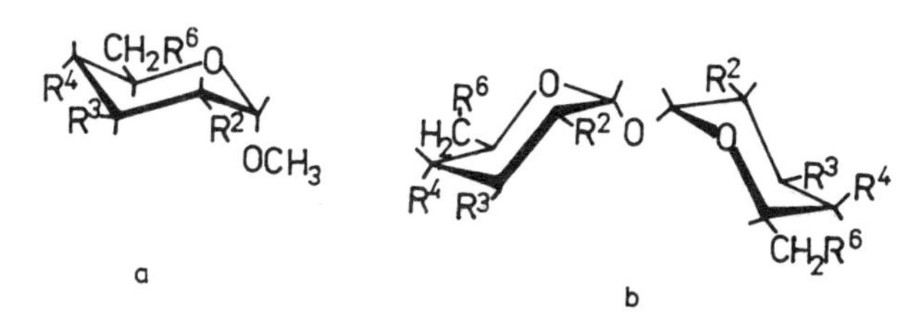

Fig. 1.9 — Sensoric properties of deoxy-*α*-D-hexopyranosides [42]. (a) 2-Deoxy-*α*-D-arabinopyranoside (R^2 = H, R^3 = R^4 = R^6 = OH), sweet with traces of bitterness; 3-deoxy-*α*-D-ribohexopyranoside (R^3 = H, R^2 = R^4 = R^6 = OH), marginally sweet; 4-deoxy-*α*-D-xylohexopyranoside (R^4 = H, R^2 = R^3 = R^6 = OH), sweet; 6-deoxy-*α*-D-glucopyranoside (R^6 = H, R^2 = R^3 = R^4 = OH), sweet with traces of bitterness. (b) 2,6-Dideoxy-*α*-D-arabinohexopyranoside (R^2 = R^6 = H, R^3 = R^4 = OH), non-sweet (very bitter); 4,6-dideoxy-*α*-D-ribohexopyranoside (R^3 = R^6 = H, R^2 = R^4 = OH), non-sweet (very bitter); 4,6-dideoxy-*α*-D-xylohexopyranoside (R^4 = R^6 = H, R^2 = R^3 = OH), non-sweet (very bitter); 2,3-dideoxy-*α*-D-glucopyranoside (R^2 = R^3 = H, R^4 = R^6 = OH), non-sweet (very bitter).

tives tasted bitter-sweet. Removal of further oxygen atoms, as in dideoxy derivatives of methyl-α-D-glucopyranoside and in the corresponding tetradeoxy derivatives of trehalose resulted in complete bitterness (Fig. 1.9).

2,6-Dideoxy derivatives taste differently in comparison with their monodeoxy analogues, and thus must align themselves differently on the receptor site.

Monitoring of sweetness of the monomethyl ethers of the two above-mentioned sugar models revealed that sweetness was not affected in the cases when the methyl group resided in positions 2, 3, 4 or 6 [83]. The analysis of these results, in conjunction with those for the dideoxy derivatives, appears to imply that the position 3 represents the B component of the AH–B system. The C_4–OH serves as the AH unit, the C_2–OH does so rarely. An interesting feature of the AH–B system in the glucopyranoside structural types is the fact that it is located in the less reactive (and thus less disturbed) part of the molecule [42] and never reaches as far as the primary hydroxyl group or the anomeric centre.

There is evidence [28] to support the suggestion that an axial-equatorial or diequatorial arrangement of α-glycol units (glucophores) in molecules of saccharides is the carrier of sweet taste. Such vicinal hydroxyl groups act as an AH–B system [84, 85] with the γ-centre as the third point [50]. The latter is situated at the opposite side of the AH–B couple. It is illustrated in Fig. 1.10 [84].

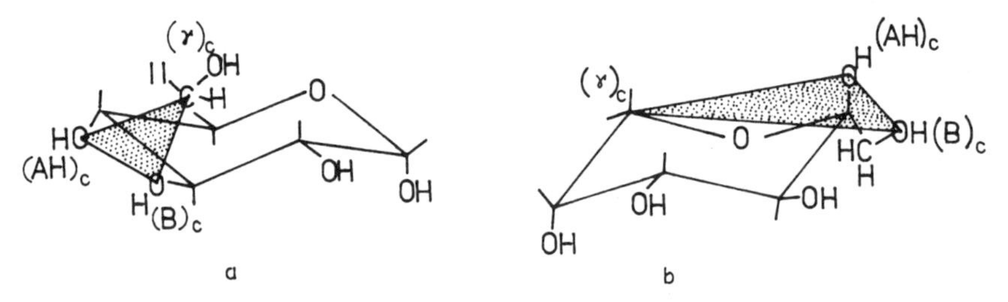

Fig. 1.10 — Position of the AH–B unit and the γ-centre of at the D-glucopyranose (a), and D-fructopyranose (b) [84].

The understanding of the structure–activity relationship (SAR) in sugars was always complicated by their mobile structures and rather unpredictable distribution of conformers in solutions. The SAR studies have been undertaken in spite of the above difficulties. The results are presented in Tables 1.4–1.6 in the form of relative sweetness values for monosaccharides, disaccharides and their derivatives, including sugar alcohols [16].

Table 1.4 — Relative sweetness of monosaccharides and their derivatives

Monosaccharide	Relative sweetness
D-Glucose	0.7
D-Mannose	0.6
D-Galactose	0.4
D-Fructose	1–1.7
6-Chloro-6-deoxy-D-glucose	traces of sweetness
6-Chloro-6-deoxy-D-galactose	0
6-Chloro-6-deoxy-D-mannose	traces of sweetness
6-Deoxy-D-glucose	sweet
6-Deoxy-D-galactose	sweet
6-Chloro-6-deoxy-D-fructofuranose	sweet
6-Chloro-1,6-dideoxy-D-fructofuranose	very sweet
D-Xylose	0.7

Table 1.5 — Relative sweetness of disaccharides and their derivatives

Disaccharide	Relative sweetness
α,α-Trehalose	sweet
6-Chloro-6-deoxytrehalose	traces of sweetness
6′,6-Dichloro-6′,6-dideoxytrehalose	traces of sweetness
4,6-Dichloro-4,6-dideoxytrehalose	0
4′,6′,4,6-Tetrachloro-4′,6′,4,6-tetradeoxygalactotrehalose	0
Sucrose	1
Lactose	0.2
Maltose	0.5
1′-Chloro-1′-deoxysucrose	20
6-Chloro-6-deoxysucrose	bitter
1′,6′-Dichloro-1′,6′-dideoxysucrose	500
6,6′-Dichloro-6,6′-dideoxysucrose	0
1′,4,6,6′-Tetrachloro-1′,4,6,6′-tetradeoxysucrose	100
1,6,6′-Trichloro-1′,6,6′-trideoxysucrose	100
Galactosucrose	0
4-Chloro-4-deoxygalactosucrose	5
1′,4-Dichloro-1′,4-dideoxygalactosucrose	600
1′,4,6′-Trichloro-1′,4,6′-trideoxygalactosucrose	2000
1′,4,6,6′-Tetrachloro-1′,4,6,6′-tetradeoxygalactosucrose	200
4,6,6′-Trichloro-4,6,6′-trideoxygalactosucrose	4
1′,4,4′,6′-Tetrachloro-1′,4,4′,6′-tetradeoxygalactosucrose	2200
6′-Chloro-6′-deoxymaltose	0
6-Chloro-6-deoxymaltose	0
6′,6-Dichloro-6′,6-dideoxymaltose	0
4′,6′-Dichloro-4′,6′-dideoxymaltose	0

Table 1.6 — Relative sweetness of sugar alcohols

Sugar alcohol	Relative sweetness
Sorbitol (Glucitol)	0.5
Mannitol	0.6
Dulcitol	0.1
Glycerol	0.8
Xylitol	1.0
Maltitol	0.6–0.9
Lycasine[a]	0.75
Palatinit[b]	0.75

[a] A product of hydrogenation of glucose — mixture of polyols.
[b] Product of hydrogenation of isomaltulose.

Surprisingly, the sweetness of sucrose, the most widely used sweetener, has been much less studied than might have been expected. Its sweetness has been assessed by stepwise blocking of the hydroxyl groups which prevents their participation in eliciting the sweet response. This is achieved by partial methylation [86] (see also p. 000). Methylation of the C'_6–OH group had zero or negligible effect on the sweet potency. On the other hand, methylation in position 4 lowered the sweetness profoundly, thus indicating that the C_4–OH participated in the receptor site interactions by hydrogen bonding. Similarly, the reduced sweetness of galactosucrose can be accounted for by the fact that its C_4–OH is blocked by an intramolecular hydrogen bond with the pyranose ring oxygen.

Sweet potency of 1′,6′-di-*O*-methylsucrose is also lower than that of sucrose. This is believed to be due to the participation of the $C_{1'}$–OH in the binding to the receptor. A favourable steric arrangement at $C_{1'}$ was discovered after the crystal structure of sucrose became known [87, 88] — it showed that both rings were held firmly in position by intramolecular hydrogen bonds.

Motivated by the unexpected discovery that 1′,4,6,6′-tetrachloro-1′,4,6,6′-tetradeoxygalactosucrose was 200 times sweeter than sucrose [89, 90], Hough and co-workers undertook the synthesis and examined structure–sweetness relationship in chlorodeoxysucrose and galactosucrose derivatives.

The extraordinary sweetness of chlorinated sucrose and galactosucrose derivatives is believed to be the consequence of the synergism of the two respective AH–B–X glucophores localized in their constituent sugar units.

The data in Table 1.5 indicate that the presence of a chlorine atom in position $C_{1'}$ is essential for the manifestation of sweet properties. It is believed that it acts as an acceptor in a hydrogen bond (B unit of a glucophore in the three-point model). If this is so, we must assign the role of an AH hydrogen donor (in a hydrogen bond with the receptor) to the equatorial 2-hydroxy group in the glucopyranose moiety. The crucial role of the third point, the lipophilic group X, must be assigned to the axial hydrogen or chlorine at the C_4 of the preferred chair conformation of the glucopyranose unit. The glucophore represented by B = $C_{1'}$–Cl, AH = C_2–OH, X = C_4–Cl of the 1′,4-dichloro-1′,4-dideoxygalactosucrose (Fig. 1.11) fits snugly into the three-point sweet unit concept of Kier [48]. The concept accounts also for the enhanced sweetness of the 1′,6′-dichloro-1′,6′-dideoxysucrose since a molecular model showed that its AH–B–X systems came very close to the interorbital dimensions of Kier's triangle [48].

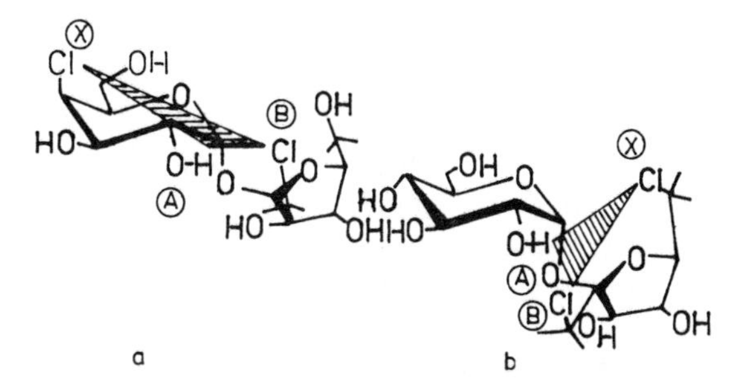

Fig. 1.11 — (a) 1′,2,4-Glucophore of 1′,4-dichloro-1′,4-dideoxygalactosucrose; (b) 1,2,6′-glucophore of 1′,6′-dichloro-1′,6′-dideoxysucrose [85].

The trichloro derivative, 1′,4′,6′-trichloro-1′,4′,6′-trideoxygalactosucrose (Fig. 1.12), is the sweetest compound in the series because of the mutual potentiation of the two glucophores [91].

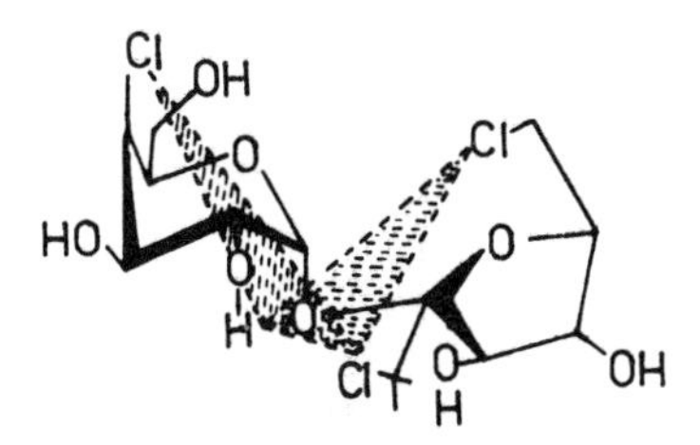

Fig. 1.12 — 1′,4,6′-Trichloro-1′,4,6′-trideoxygalactosucrose [92].

The substitution of OH by chlorine in a sugar molecule is not always beneficial for the overall sweetness, as some data in Table 1.5 indicate. Clearly, the substitution of C_6–OH of the pyranose unit by chlorine results in decreased sweetness of the deoxysucrose derivatives. The reason for this adverse effect appears to be the steric hindrance to the axial X group at C_4, which in turn prevents C_2–OH from interacting with the receptor.

1.2.6 Solution properties of saccharides *vs.* sweet taste

Although it is true that a 'glucophore' can be identified in all sweet substances, the reverse need not be universally true. In other words, molecules need not be sweet even though a glucophore has been identified in them. Clearly, a molecule needs more than a glucophore in order to be sweet. The molecule must reach the appropriate region of the sweet receptor before it can bind to it and thus activate the ion-channel mechanism. The accession efficiency is mediated by water; it is probable that no molecule can be tasted without being first dissolved in water. Hence, a study of solution properties, a factor with a direct bearing on sweet properties, has been proposed. Some authors [93] approached the problem by correlating the solute–solvent interactions with physical properties of aqueous solutions of small sugar molecules — and the resultant taste [94]. Some idea of the importance of water in the process of eliciting sweet taste can be gleaned from the results of electrophysiological experiments carried out in Japan [95]. Indeed, the adsorption of the sweet molecule on the receptor site induces a potential in the taste cells, probably due to ion transfer across the membrane.

The fundamental water–sugar interaction problem can be formulated within three brief postulates:

(a) mono- and disaccharides exist in aqueous solution in conformational and configurational equilibria
(b) weak energy solute–solvent interactions enable specific solvation (hydration) of each isomer
(c) the arrangement of the hydrophilic and hydrophobic groups around the heterocyclic ring of sugars affects the 'water structure' in the immediate vicinity of sugar molecules

The understanding of conformational and configurational equilibria of sugars in aqueous solutions forms an important prerequisite and an integral part of a more general problem of correlating structure with sweetness. The problem can be illustrated by some examples. D-Fructose, β-D-fructopyranose in its crystalline state, exists in an aqueous solution as a complex mixture in dynamic equilibrium comprising not only of the β-D-pyranose anomer but also β-D-fructofuranose, α-D-fructofuranose and α-D-fructopyranose [85, 96]. The relative proportion of the respective isomers is governed by the solvent properties — thus in pyridine even the open chain form of D-fructose was detected [97]. Another well-known example is the mutarotation of D-glucose in water which leads to the establishment of an equilibrium containing 38 % of the α-anomer and 62 % of the β-anomer [98], furanose or acyclic forms have not been detected. As in the previous example, the position of the equilibrium depends on the solvent [99].

The ^{13}C NMR studies of sucrose in dilute deuterated solutions [100] have provided evidence for intramolecular hydrogen bonding. The effect of solvent on both intra- and intermolecular hydrogen bonding have been examined [101] and resulted in the formulation of the concept of conjugated hydrogen bonds. The concept was elaborated further by Jeffrey and Takagi [102].

The interactions between various solvents and sugars can be assessed by various experimental techniques [103–105]. However, as is often the case, the results depend greatly on the technique used.

The first layer of solvating water molecules is in the most intimate contact with the saccharide molecule and thus affects most the properties of the sugar hydroxyl groups. Because of the low energy of such encounters, the hydration is highly specific and depends both on conformation and configuration of the sugar. The most significant consequence of the hydration is the fact that water molecules compete for partners (at the receptor site) in hydrogen bonding with the AH–B system of sugars, and thus must be considered as an integral part of the whole picture [93].

Ryazanov [106] recognizes layers in the arrangement of solvating water molecules and defines the short-range order around the sugar molecule (the innermost water layer) as *hydration* and the more distant layer, which is the long-range order of the bulk water, as *water structure*. Semiempirical analysis of laser Raman spectra of water [107] uncovered several (at least five) types of association in water. When this technique was applied to aqueous solutions of sugars, it was found [93] that traces of D-glucose, D-fructose or sucrose modified the proportion of the species of various association types between water molecules.

The experimental technique in this area of research has tended to gravitate towards a combination of several techniques [108–114]. The most common method for learning about the hydrodynamic behaviour of molecules in dilute solutions is based on the determination of viscometric constants, the intrinsic viscosity and Higgins constants [116], apparent molar and specific volumes and other related properties.

Infrared and Raman spectroscopy, which are able to detect free (unassociated) hydroxyl groups, serve as a tool for classifying saccharides as either promoters or breakers of water structure [115].

Apparent molar volumes of sugars depend on the distribution of axial and equatorial hydroxyl groups [117]. The reason for this is that equatorial hydroxyl groups hydrate more easily than axial ones. Better hydrated structures are also more compatible with water structure and therefore show lower apparent molar volumes. For instance, if we compare β-D-glucopyranose with all its hydroxyl groups equatorial and its α-anomer, we see that both their apparent molar volumes [113] (Table 1.7) and their sweetness are about the same [118]. On the other

Table 1.7 — Apparent molar volumes (Φ_V) of axially and equatorially substituted pyranoses [113]

Position of group	Φ_V (cm^3/mol)[a] axial	equatorial	$\Delta\Phi_V$
C_1	110.8	110.8	0
	α-D-glucose	β-D-glucose	
C_2	110.8	110.8	0
	D-mannose	D-glucose	
	93.8	93.8	0
	D-xylose	D-xylose	
C_3	94.78	93.80	0.98
	D-ribopyranose	D-xylose	
C_4	109.0	110.8	1.80
	D-galactose	D-glucose	
	91.84	93.80	1.96
	L-arabinose	D-xylose	

[a] 3% concentration.

hand, D-galactopyranose (an axial C_4–OH) is less sweet than D-glucopyranose (equatorial C_4–OH). Consequently, it is not surprising that α-D-mannopyranose (axial C_2–OH) and D-glucopyranose are nearly equally sweet as both molecules possess the same apparent molar volume. The differences in apparent molar volumes Φ_V of axial and

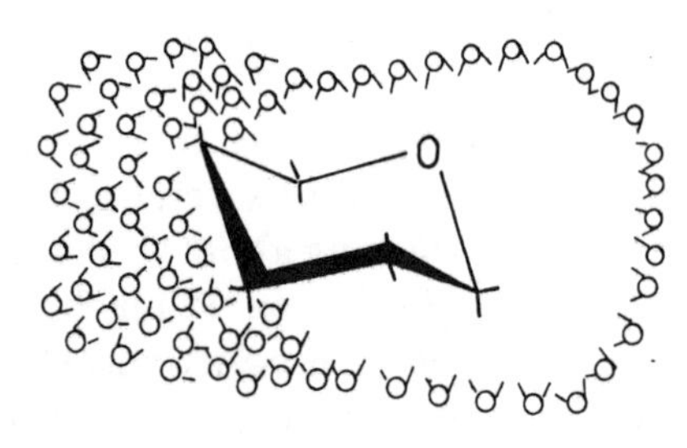

Fig. 1.13 — Postulated interaction of water molecules around sugar molecule [113].

equatorial isomers of pyranoses is a consequence of the predominant hydration of hydrophilic parts of the sugar molecule (Fig. 1.13) [113].

The delicate interrelation of axial-equatorial conformational changes at C_3 and C_4 which induce changes in the water structure, and in Φ_V values, is clearly evident from the data in Table 1.7. Apparent molar volumes thus form a link between the sweetness and the structural elements of C_4 atoms and their axial-equatorial hydroxyl groups. Therefore, an alteration in the 3,4-diequatorial system of D-glucose not only disturbs the surrounding water structure, but also decreases the intensity of sweetness. The present state of knowledge in this area has not so far allowed predictions of the duration of sweet sensation which are based on the apparent molar volumes.

However, it can be said that in a mixture of two sugars, the one with lower apparent molar volume will dominate in the sweet sensation [109, 119].

It appears from the studies of apparent specific volumes [110] that a majority of pleasantly tasting substances can be fitted into a general pattern related to taste (Table 1.8).

Table 1.8 — Differentiation of basic taste by apparent specific volumes [110]

Apparent molar volume Φ_V/m.wt. (cm^3/g)	Taste region
0–0.33	salty
0.33–0.52	sour
0.52–0.71	sweet
0.71–0.90	bitter

Compounds that have an apparent specific volume greater than 0.93 cm^3/g are probably too hydrophobic to be perceptible to human taste organs. Furthermore, it is evident from Table 1.8 that compounds with different apparent specific volumes elicit different responses at the taste receptor, thus reflecting the efficiency of access of the solute to the receptor.

The difference in sweetness between D-glucose and D-fructose is also manifested in the infrared spectra. Whilst only one three-point glucophore has been suggested (and accepted) for β-D-fructopyranose, Mathlouthi and Seuvre [114] stated two pairs of AH–B hydrogen bonds for D-glucose (Fig. 1.14).

The unique character of AH–B groups manifests itself by the presence of a free hydroxyl group and by the fact that β-pyranose form prevails in aqueous solution. The two compounds in question also differ in their respective hydrophobic parts (γ-centre); it is, in the case of fructose, a methylene group which is more hydrophobic than the CH_2–OH group, the γ-centre of D-glucose. It was suggested [111] that the γ-centre is responsible for the increased sweetness of D-glucose in comparison with D-galactose.

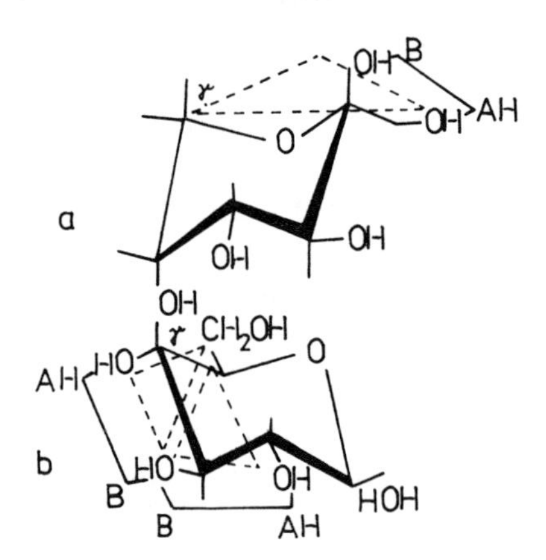

Fig. 1.14 — Postulated glucophores of β-fructopyranose (a), and D-glucopyranose (b) [114].

The very sweet 1′,4,6′-trichloro-1′,4,6′-trideoxygalactosucrose was already discussed and it was concluded that its high sweetness was probably due to the peculiar arrangement of chlorine atoms which resulted in an AH–B system and a hydrophobic centre located at the opposite sides of the sugar molecule. Figure 1.15 illustrates the struc-

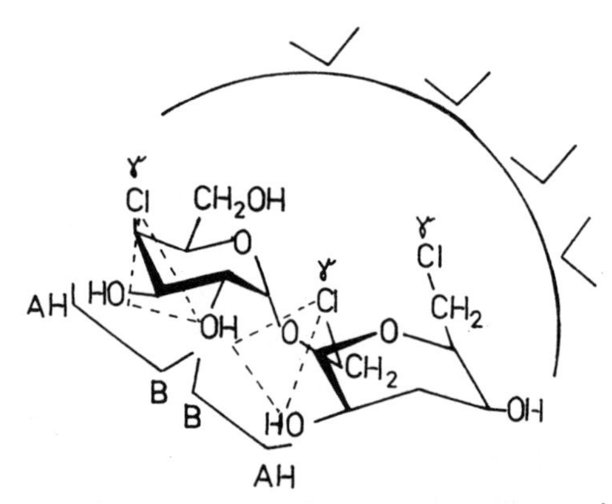

Fig. 1.15 — An interaction that water is believed to undergo with the hydrophobic centre of 1′,4,6′-trichloro-1′,4,6′-trideoxygalactosucrose. Water molecules (v) are repelled by the hydrophobic side.

ture of the glucophore as well as its hydrophobic interaction with water [114].

The mobility of water molecules in the immediate vicinity of the sweet solute plays a very important role in the perception of taste, as it can enhance it. This is the reason for D-fructose being sweeter than most natural saccharides. A change in the isoosmotic equilibrium at both sides of the receptor membrane results in a more rapid Na^+/K^+ transfer and hence a more intense sweet response.

A great deal of understanding has been gathered on the interaction of saccharides with solvents; this knowledge, in combination with the so far developed structure–sweetness relationship, accounts quite satisfactorily for the observed variations in the potency of sweetness. However, in spite of this success, further progress in this area is being complicated by additional new factors not considered to date. One such factor is the temperature dependence of the sweet sensation [120, 121]. Hence, sweet taste must be characterized as a complex sensation which involves an interplay of structural and solution properties of the sweetener. Better understanding of this delicate interplay will enable us to learn more about the chemoreception of sweet taste as well.

REFERENCES

1. Paré, J. R. J., Milon, H. & Bélanger, J. M. R. (1988) *Canad. Chem. News* 13.
2. Kawamura, Y. & Kare, M. R. (eds) (1987) *Umami: A Basic Taste*. M. Dekker, New York.
3. Pevzner, P. A. (1978) In: Gushuni, G. V. (ed.) *Sensoric Systems*. Nauka, Moscow, p. 115 (in Russian).
4. Kurihara, K. & Beidler, L. M. (1969) *Nature* **222**, 1176.
5. Bratus, T. N., Kozlova, M. V. & Lebedeva, V. A. (1978) In: Gushuni, G. V. (ed.) *Sensoric Systems*. Nauka, Moscow, p. 138 (in Russian).
6. Dastoli, F. R. & Price, S. (1966) *Science* **154**, 905.
7. Hiji, Y., Kobayashi, N. & Sato, M. (1968) *Kumamoto Med. J.* **21**, 137.
8. Hiji, Y., Kobayashi, N. & Sato, M. (1971) *Comp. Biochem. Physiol.* **39**, 367.
9. Sato, M., Hiji, Y. & Ito, H. (1977) In: Le Maguen, & MacLeod, P. (eds) *Olfaction and Taste VI*. Information Retrieval, London, p. 233.
10. Sato, M., Hiji, Y. & Imoto, T. (1971) In: Kare, R. M. & Maller, O. (eds) *The Chemical Senses and Nutrition*. Academic Press, New York, p. 327.
11. Sato, M. (1985) *Jpn. J. Physiol.* **35**, 875.
12. Beidler, L. M. (1954) *J. Gen. Physiol.* **38**, 133.
13. Beidler, L. M. (1966a) *J. Food Sci.* **31**, 271.
14. Beidler, L. M. (1966b) *Adv. Chem.* **56**, 1.
15. Cagan, R. H. (1971) *Biochim. Biophys. Acta* **252**, 199.
16. van der Wel, H., van der Heijden, A. & Peer, H. G. (1987) *Food Rev. Int.* **3**, 193.
17. Brouwer, J. N., Hellekant, G., Kasahara, Y., van der Wel, H. & Zotterman, Y. (1973) *Acta Physiol. Scand.* **89**, 550.

18. Hellekant, G., Brouwer, J. N., Glaser, D. & van der Wel, H. (1976) *Acta Physiol. Scand.* **97**, 241.
19. Glaser, D., Hellekant, G., Brouwer, J. N. & van der Wel, H. (1978) *Folia Primatol.* **29**, 56.
20. Naim, M., Rogatha, H., Yamamoto, T. & Zehari, V. (1982) *Physiol. Behav.* **28**, 979.
21. van der Wel, H. & Arvidson, K. (1978) *Chem. Senses* **3**, 291.
22. Pritchard, T. C. & Scott, T. R. (1982) *Brain Res.* **253**, 81, 93.
23. Faurion, A., Bonaventure, L., Bertrand, B. & Mac Leod, P. (1980) In: van der Starre, H. (ed.) *Olfaction and Taste VII.* Information Retrieval, London, p. 86.
24. Zawalich, W. S. (1972) In: Schneider, D. (ed.) *Olfaction and Taste IV.* Wissenschaftliche Verlagsgesellschaft, Stuttgart, p. 280.
25. Faurion, A., Saito, S. & Mac Leod, P. (1980) *Chem. Senses* **5**, 107.
26. Schiffman, S. S., Lindley, M. G., Clark, I. B. & Makimo, H. (1981) *Neurobiol. Aging* **2**, 173.
27. Schiffman, S. S., Cahn, H. & Lindley, M. G. (1981) *Pharmacol. Biochem. Behav.* **15**, 377.
28. Birch, G. G. (1976) *Crit. Rev. Food Sci. Nutr.* **8**, 57.
29. Oertly, E. & Myers, R. G. (1919) *J. Am. Chem. Soc.* **41**, 855.
30. Cohn, G. (1914) *Pharm. Zentralhalle Dtsch.* **55**, 735.
31. Kodama, S. (1920) *J. Tokyo Chem. Soc.* **41**, 495.
32. Tsuzuki, Y. (1948) *Chem. Ind.* (Japan) **1**, 32.
33. Deutsche, E. W. & Hansch, C. (1966) *Nature* **211**, 75.
34. Steinhardt, R. G., Calcin, A. D. & Dodd, E. A. (1962) *Science* **135**, 367.
35. Verkade, P. E. (1968) *Farmaco, Ed. Sci.* **23**, 248.
36. Shallenberger, R. S. (1963) *J. Food Sci.* **28**, 584.
37. Shallenberger, R. S. (1964a) *Agric. Sci. Rev.* **2**, 11.
38. Shallenberger, R. S. (1964b) *New Sci.* **407**, 569.
39. Shallenberger, R. S., Acree, T. E. & Guild, W. E. (1965) *J. Food Sci.* **30**, 560.
40. Shallenberger, R. S. & Acree, T. E. (1967) *Nature* **216**, 480.
41. Shallenberger, R. S. & Acree, T. E. (1969a) *J. Agric. Food Chem.* **17**, 701.
42. Birch, G. G. & Lee, C. K. (1979) *Dev. Sweeteners* **1**, 165.
43. Birch, G. G., Cowell, N. D. & Eytow, D. (1970) *J. Food Technol.* **5**, 277.
44. Shallenberger, R. S. & Acree, T. E. (1969b) *Nature* **221**, 555.
45. Solms, J. (1969) *J. Agric. Food Chem.* **17**, 686.
46. Belitz, H. D. & Wieser, H. (1976) *Z. Lebensm.-Unters. Forsch.* **160**, 251.
47. Wieser, H., Jugel, H. & Belitz, H. D. (1977) *Z. Lebensm.-Unters. Forsch.* **164**, 277.
48. Kier, L. B. (1972) *J. Pharm. Sci.* **61**, 1394.
49. Shallenberger, R. S. (1977) In: Birch, G. C., Brennan, J. G. & Patker, K. S. (eds), *Sensory Properties of Foods.* Appl. Sci. Publ. Ltd., London, p. 91.
50. Shallenberger, R. S. & Lindley, M. G. (1977) *Food Chem.* **2**, 145.
51. Shallenberger, R. S. (1980) *Food Technol.* (Chicago) **34**, 65.
52. Ferguson, L. N. & Lawrence, A. R. (1958) *J. Chem. Educ.* **35**, 436.
53. Lawrence, A. R. & Ferguson, L. N. (1959) *Nature* **183**, 1469.
54. Barnes, C. J. & Ferguson, L. N. (1960) *Nature* **186**, 617.
55. Lawrence, A. R. & Ferguson, L. N. (1960) *J. Org. Chem.* **25**, 1220.
56. Ferguson, L. N. & Childers, L. G. (1960) *J. Org. Chem.* **25**, 1921.
57. Hansch, C., Steward, A. R. & Iwasa, J. (1965) *Mol. Pharmacol.* 1.
58. Blanksma, J. J. & Hoegen, D. (1946) *Rec. Trav. Chim.* **65**, 333.
59. Kier, L. B. (1974) In: Benz, G. (ed.) *Structure–Activity Relationship in Chemoreception.* Information Retrieval, London, p. 101.

60. Höltje, H. D. & Kier, B. (1974) *J. Pharm. Sci.* **63**, 1722.
61. Kier, L. B., Hall, L. H., Murray, W. J. & Randle, M. (1975) *J. Pharm. Sci.* **64**, 1971.
62. Murray, W. J. & Kier, L. B. (1976) *J. Med. Chem.* **19**, 573.
63. Kier, L. B. & Hall, L. H. (1976) *J. Pharm. Sci.* **65**, 1806.
64. Verloop, A., Hoogstraaten, W. & Tripker, J. (1976) In: Ariens, E. J. (ed.) *Drug Design,* Vol. 3. Academic Press, New York, p. 165.
65. Iwamura, H. (1980) *J. Med. Chem.* **23**, 308.
66. Ferguson, L. N. (1981) *J. Chem. Educ.* **58**, 456.
67. Tinti, J. M., Durozard, D. & Nofre, C. (1980) *Naturwissenschaften* **67**, 193.
68. Tinti, J. M. & Nofre, C. (1981) *Naturwissenschaften* **68**, 143.
69. Tinti, J. M., Nofre, C. & Peytavi, A. M. (1982) *Z. Lebensm.-Unters. Forsch.* **175**, 266.
70. Pavlova, L. A., Komarova, T. V., Davidovich, Yu. A. & Rogozhin, S. V. (1981) *Usp. Khim.* **4**, 590 (in Russian).
71. Temussi, P. A., Lelj, F. & Tancredi, T. (1978) *J. Med. Chem.* **21**, 1154.
72. Lelj, F., Tancredi, T., Temussi, P. A. & Toniolo, C. (1976) *J. Am. Chem. Sci.* **98**, 6669.
73. Lelj, F., Tancredi, T., Temussi, P. A. & Toniolo, C. (1980) *Farmaco, Ed. Sci.* **35**, 988.
74. Temussi, P. A., Lelj, F., Tancredi, T., Castiglione-Morelli, M. A. & Pastore, A. (1984) *Int. J. Quantum Chem.* **26**, 889.
75. Hopfinger, A. J. & Jabloner, H. (1981) In: Charalambous, G. & Inglett, G. (eds) *The Quality of Food and Beverages.* Academic Press, New York, p. 83.
76. Venkatachalam, C. M. (1968) *Biopolymers* **6**, 1425.
77. Acton, E. M. & Stone, H. (1968) *Science* **193**, 584.
78. Hansch, C. & Leo, A. (1979) In: *Substituent Constants for Correlation Analysis in Chemistry and Biology.* J. Wiley, New York, p. 352.
79. Iwamura, H. (1981) *J. Med. Chem.* **24**, 572.
80. van der Heijden, A., van der Wel, H. & Peer, H. G. (1985) *Chem. Senses* **10**, 57.
81. Birch, G. G., Lee, C. K. & Rolfe, E. J. (1970) *J. Sci. Food Agric.* **21**, 650.
82. Birch, G. G. & Lee, C. K. (1974) *J. Food Sci.* **39**, 947.
83. Lindley, M. G. & Birch, G. G. (1975) *J. Sci. Food Agric.* **26**, 117.
84. Lindley, M. G., Birch, G. G. & Khan, R. (1976) *J. Sci. Food Agric.* **27**, 140.
85. Shallenberger, R. S. (1973) *Adv. Chem. Ser.* **117**, 256.
86. Lindley, M. G., Birch, G. G. & Khan, R. (1976) *Carbohydr. Res.* **43**, 140.
87. Brown, G. M. & Levy, H. A. (1963) *Science* **19**, 921.
88. Brown, G. M. & Levy, H. A. (1973) *Acta Crystallogr., B* **29**, 790.
89. Hough, L. & Phadnis, S. P. (1976) *Nature* **263**, 800.
90. Hough, L. & Khan, R. (1978) *Trends Biochem. Sci.* **3**, 61.
91. Khan, R. (1979) In: Birch, G. G. & Parker, K. I. (eds) *Sugar: Science and Technology.* Appl. Sci. Publ. Ltd., London, p. 181.
92. Khan, R. (1979) *J. Chem. Soc. Pak.* **1**, 34.
93. Mathlouthi, M. (1984) *Food Chem.* **13**, 1.
94. Shamil, S., Birch, G. G., Mathlouthi, M. & Clifford, M. N. (1987) *Chem. Senses* **12**, 397.
95. Kobatake, Y., Kurihara, K. & Kamo, N. (1978) In: Takamura, T. & Kozawa, A. (eds), *Surface Electrochemistry.* Japan Scientific Societies Press, Tokyo, p. 1.
96. Doddrell, D. & Allerhand, A. (1971) *J. Am. Chem. Soc.* **93**, 2.
97. Funcke, W. & Klemer, A. (1976) *Carbohydr. Res.* **50**, 9.
98. Isbell, H. S. & Pigman, W. W. (1938) *J. Res. Natl. Bur. Stand., A* **20**, 773.
99. Franks, F. (1979) In: Blanchard, J. M. & Mitchell, J. R. (eds) *Polysaccharides in*

Foods. Butterworths, London, p. 33.
100. Bok, K. & Lemieux, R. U. (1982) *Carbohydr. Res.* **100**, 63.
101. Lemieux, R. U. & Pavia, A. A. (1969) *Can. J. Chem.* **47**, 4441.
102. Jeffrey, G. A. & Takagi, S. T. (1978) *Acc. Chem. Res.* **11**, 264.
103. Robinson, R. A. & Stockes, R. H. (1966) *J. Phys. Chem.* **70**, 2126.
104. Harvey, J. M., Symons, M. C. R. & Naftalin, R. J. (1976) *Nature* **261**, 435.
105. Bociek, S. & Franks, F. (1979) *J. Chem. Soc., Faraday Trans. 1* **2**, 262.
106. Ryazanov, M. A. (1978) *Russ. J. Phys. Chem.* **52**, 754.
107. Luu, C., Luu, D. V. & Sopron, F. (1982) *J. Mol. Struct.* **81**, 1.
108. Birch, G. G. & Catsoulis, S. (1985) *Chem. Senses* **10**, 325.
109. Birch, G. G. & Shamil, S. (1986) *Food Chem.* **21**, 245.
110. Shamil, S., Birch, G. G., Mathlouthi, M. & Clifford, M. N. (1987) *Chem. Senses* **12**, 397.
111. Mathlouthi, M., Seuvre, A. M. & Birch, G. G. (1986) *Carbohydr. Res.* **152**, 47.
112. Shamil, S., Birch, G. G. & Njoroge, S. (1988) *Chem. Senses* **13**, 457.
113. Birch, G. G. & Shamil, S. (1988) *J. Chem. Soc., Faraday Trans. 1* **84**, 2635.
114. Mathlouthi, M. & Seuvre, A. M. (1988) *J. Chem. Soc., Faraday Trans. 1* **84**, 2641.
115. Jackson, S. E. & Symons, M. C. R. (1976) *Chem. Phys. Lett.* **37**, 551.
116. Huggins, M. (1942) *J. Am. Chem. Soc.* **64**, 2716.
117. Franks, F., Ravenhill, J. R. & Reid, D. S. (1972) *J. Solution Chem.* **1**, 3.
118. Birch, G. G., Shamil, S. & Shepherd, Z. (1986) *Experientia* **42**, 1232.
119. Munton, S. L. & Birch, G. G. (1985) *J. Theor. Biol.* **112**, 539.
120. Green, B. G. & Frankmann, S. P. (1987) *Chem. Senses* **12**, 609.
121. Green, B. G. & Frankmann, S. P. (1988) *Physiol. Behav.* **43**, 515.

2

Naturally occurring sweet substances

Sugar has been always an essential part of many foods, either as a naturally occurring compound or as an added ingredient. It is probable that the early hominids had a 'sweet tooth' and enriched their diet with sweet tasting components, and consequently had to endure various hardships such as long travels, painful bee stings and even fights with animals. The oldest available naturally occurring edible sweet matter was probably honey and fruit and plant juices. The dominant sweetener of the present — sucrose (saccharose, sugar) — was first isolated from sugar cane, a plant that is generally believed to have originated in Oceania, from where it spread to India. Europeans learned about it during the campaigns of Alexander the Great in 327 B.C. His commanders described a plant that produced "honey without the help from bees and gave a sweet drink, even though it bore no fruit". Small scale sugar cane growing started in Europe in the 8th century and sugar became a commercial product. The bulk of sugar that was needed had to be imported because the alternative domestic sources of sweeteners, made from maple and birch syrups, from carrots and other plants, could not supply the demand.

Sugar was considered for a long time as a luxury ingredient for sweet drinks and confectionery, but it was difficult to obtain. The demand for sugar increased rapidly when the consumption of beverages such as tea, coffee and chocolate became widespread.

Ultimately, Marggraf succeeded in isolating sucrose from the red beet in the 18th century. His pupil, Archard, later increased the yield of isolated sucrose from sugar beet so that sugar could be produced from it industrially. Sugar industry received a big impetus during the Napoleonic wars when the British blockade cut off imports to France. Half a century after Napoleon's decree of 1806, which was designed to encourage farmers to grow more sugar beet, the production of sugar from this feedstock increased from the original 35 tonnes/year to 45 000 tonnes in 1855.

Sugar (sucrose) has been for a long time the only safe natural sweetener in the subconscious mind of the consumer. Indeed, this has narrowed the meaning of the term 'sugar' solely to sucrose for all but the chemist. Owing to the relative stability of sucrose in aqueous solutions, its pleasant and pure taste became a standard of sweet taste quality against which all other 'surrogate' sweeteners are compared.

2.1 SACCHARIDES

Saccharides, often called sugars or glycogens, occur naturally in plants and animals, where they are the source of chemical energy and as the constituent of the supporting tissues in the plants. Sucrose is found, for instance, in the roots and stems of all grasses, in various root vegetables such as parsnip, carrot, turnip and beet and in the sap of many trees, especially sugar maple and many palm trees.

Saccharides [1] are, according to the most common definition, polyhydroxy aldehydes and polyhydroxy ketones. Although this definition stresses the principal functional groups of sugars, it is not quite adequate because it disregards the fact that the most often encountered saccharide molecule is in the cyclized hemiacetal or acetal form.

The simplest saccharides that cannot be cleaved hydrolytically into smaller saccharide units are called *monosaccharides*. The term *oligosaccharides* comprises a family of saccharides that contain between two to ten monosaccharide units. Saccharides constructed from many monosaccharide units are called *polysaccharides*.

2.1.1 Sucrose (sugar)

Common sugar (from sugar beet or sugar cane) — sucrose — is a non-reducing disaccharide (Fig. 2.1). Its systematic name is β-D-fructofuranosyl-α-D-glycopyranoside. Glucose carbon atoms are conventionally numbered, those of fructose are primed. Facts about sucrose, which has 8 OH groups and the α-D-glucopyranosyl and β-D-fructofuranoside residues in $^{4}C_{1}$ and $_{3}T^{4}$ conformations, respectively, fill several compendia [2–4].

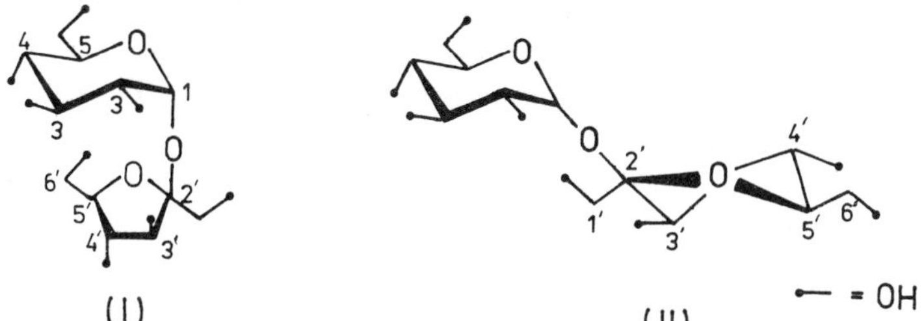

Fig. 2.1 — Sucrose (β-D-fructofuranosyl-α-D-glucopyranoside).

Molecular structure of sucrose has been confirmed by enzymatic [5–7] and chemical [8–11] synthesis, X-ray analysis [12, 13] and by NMR spectroscopy [14]. Brown and Levy [15] determined the exact crystal and molecular structures of sucrose by neutron diffraction method. They showed that in anhydrous crystalline sucrose seven hydroxyl groups participate in hydrogen bonds, two of which are intermolecular (O-2...HO-1′ and O-5...HO-6′) (III) (Fig. 2.2). ^{1}H and

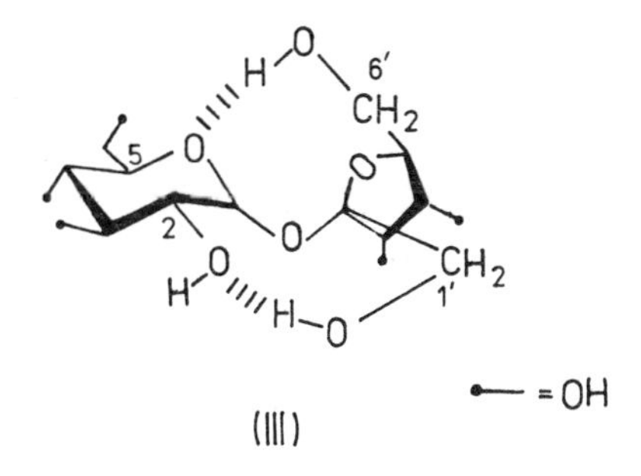

Fig. 2.2 — Intramolecular hydrogen bonds in the molecule of sucrose.

^{13}C NMR [16] was used to confirm that sucrose, even when dissolved in water, retained this conformation. In contrast, others have found an equilibrium of two conformations in $(CD_3)_2SO$ solution [17], which is characterized by two types of hydrogen bonds in which HO-2 acts as an acceptor for HO-1′ or HO-3′ of the fructofuranosyl moiety. The relative concentration of the two conformations in equilibrium was found to be 2 : 1. Further study showed that the behaviour of sucrose in oligosaccharides remained essentially the same [18].

2.1.1.1 Derivatives of sucrose

An examination of the minute differences in reactivity of the 8 hydroxyl groups of sucrose were carried out in order to attempt selective substitution. Although the theoretical number of possible isomers is very large indeed, only a small fraction of them could be made (Table 2.1).

The explanation of this fact is simple — the primary OH groups are considerably more reactive than the secondary ones. Thus the tosylation of sucrose yields predominantly 6,1′,6′-tri-*O*-tosylate, 6,6′-di-*O*-tosylate and 2,6,6′-tri-*O*-tosylate [19]. Similar results were obtained in the tritylation reaction with trityl chloride [20]. These results enabled the classification of the hydroxyl groups in sucrose in order of their reactivity: HO-6, HO-6′ > HO-1′ > HO-2.

Table 2.1 — Number of possible isomers of sucrose derivatives

Substitution	Number of isomers	Substitution	Number of isomers
mono	8	penta	56
di	28	hexa	28
tri	56	hepta	8
tetra	70	octa	1

Tosylated sucroses are useful starting materials for the synthesis of other mono-, di- and trisubstituted compounds with azido [21], amino [21–23], halogeno and other functionalized groups [24].

Several methylation procedures are known which yield methyl ethers of sucrose, such as alkaline methylation with dimethyl sulphate [25, 26], treatment with iodomethane and silver oxide in acetone, sodium hydride and iodomethane in dimethylformamide [26] and methylation with diazomethane and boron trifluoride in ether [27, 28].

The last method is particularly suitable for the methylation of partially esterified derivatives of sucrose as it proceeds under mild conditions and thus does not catalyse ester group migrations. Methylated sucroses have often been used as model compounds in the elucidation of the mechanism of action of sucrose on sweet receptors [28] (Table 2.2).

Table 2.2 — Sweetness of methylated sucroses

Compound	Sweetness
Sucrose	very sweet
6′-*O*-Methylsucrose	very sweet
4-*O*-Methylsucrose	sweet
6,6′-di-*O*-Methylsucrose	very sweet
4,6′-di-*O*-Methylsucrose	sweet
4,6-di-*O*-Methylsucrose	sweet
1′,6′-di-*O*-Methylsucrose	sweet

The chlorodeoxy derivatives of sucrose are probably the most important because they are intensely sweet. As could be expected, a very large number of variously substituted derivatives have been prepared and their sweetness studied [29–32].

Although the first paper on the exchange of OH group by chlorine was published as early as 1921 [33], the real revival of interest in these compounds started four decades later when the reports by Jones and his group were published [34, 35]. The application of the substitution reactions to sucrose produced [36], as might be expected, extremely complex mixtures. Sucrose could be chlorinated with sulphuryl chlorine under controlled conditions to give 6′-chlorosucrose (IV) (Fig. 2.3),

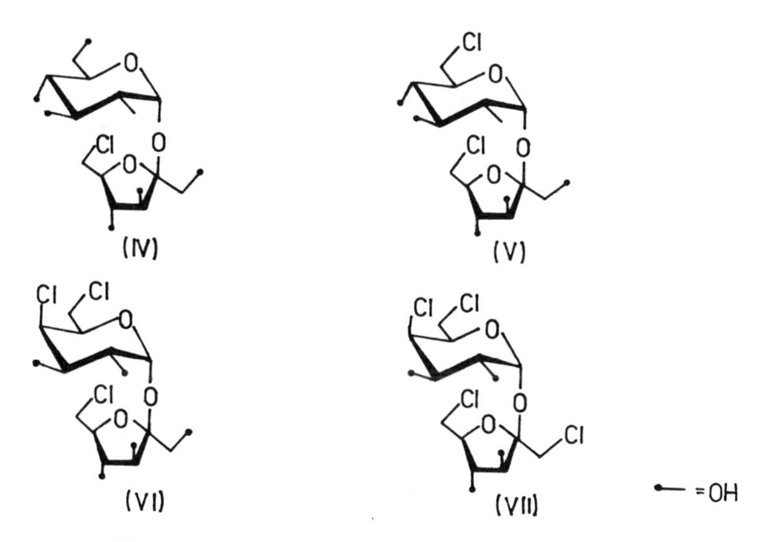

Fig. 2.3 — Chlorinated sucroses and galactosucroses.

6,6′-dichlorosucrose (V) [37], 4,6,6′-trichlorogalactosucrose (VI) [38, 39] and 1′,4,6,6′-tetrachloro derivative of galactosucrose (VII) [4]. The reaction was stereoselective on account of the mild conditions. The reactivity of the hydroxyl groups decreased in the order OH-6′ > OH-6 > OH-4 > OH-1′, OH-4′. Surprisingly, OH-4 was more reactive than the primary OH-1′. The substitution at C_4 proceeds with inversion of configuration, a change from gluco to galacto series. 6,6-Dichlorosucrose can be prepared by a more suitable and higher yield method which uses triphenylphosphine and tetrachloromethane in pyridine [40].

The use of sulphuryl chloride under harsher conditions may yield even the 1′,4,4′,6,6′-pentachloro derivative (VIII), which can be transformed into 3,6;3′,6′;2,1′-trianhydro derivative (IX) (Fig. 2.4) by treat-

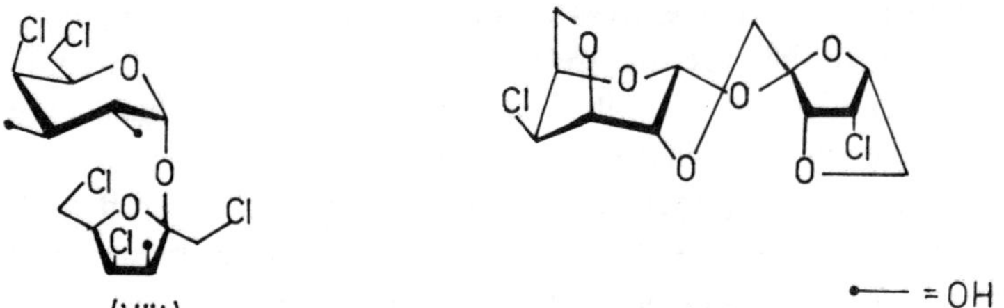

Fig. 2.4 — Derivatives of sucrose.

ment with base. Other chlorinated sucroses undergo similar transformations [4].

An interesting sequence of reactions, that leads ultimately to a sucrose derivative which is 100 times sweeter, starts with 1′,6,6′-trichlorosucrose (XI) which is made from 1′,6,6′-trimesitylenesulphonate (X) and LiCl. Further chlorination of (XI) with sulphuryl chloride gives the expected 1′,4,6,6′-tetrachlorogalactosucrose (‘serendipitose’) (VII) [41] (Scheme 2.1).

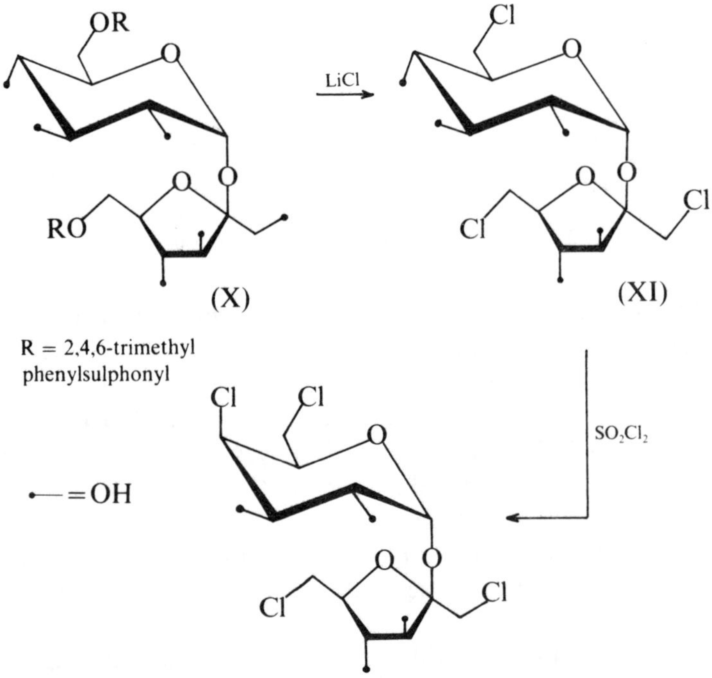

R = 2,4,6-trimethyl phenylsulphonyl

Scheme 2.1

Establishing the reason for the high sweetness of serendipitose relative to its structure was particularly revealing because the result was not expected on the basis of extrapolation from the properties of related compounds. Not only galactosucrose itself is not sweet but the 4,4′,6,6′-tetrachloro-4,4′,6,6′-tetradeoxygalactotrehalose (XIII) (Fig. 2.5) is bitter like quinine [42]. This demonstrates yet again the importance of the C_4 of sucrose for the manifestation of sweet taste.

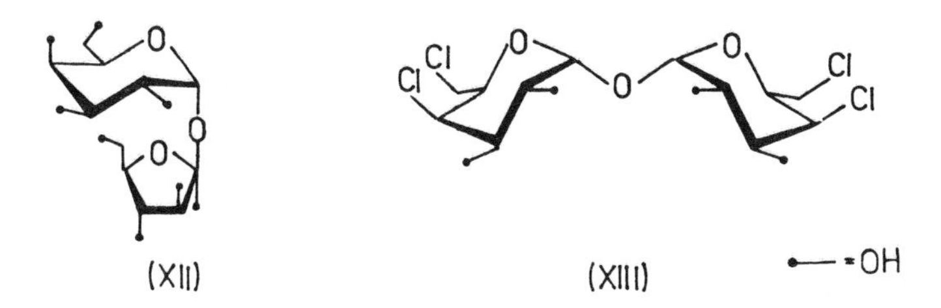

Fig. 2.5 — Galactosucrose (XII) and tetrachlorogalactotrehalose (XIII).

Prior to the discovery of the exceptionally sweet serendipitose, modifications of the sucrose molecule usually resulted in the decrease or even the loss of sweetness [43], for instance sucrose monoacetate [44], various benzoates which are all bitter [45, 46] and the octaacetate of sucrose which was identified as the bitter principle of *Clematis japonica* and used as a denaturing agent.

One of the sweetest representatives of the family of chlorinated sucroses [29, 30] is 1′,4,6′-trichloro-1′,4,6′-trideoxygalactosucrose (XV) which was prepared from 1′,6,6′-tri-*O*-tritylether pentaacetate (XIV) which on detritylation undergoes a 4→6 acetyl migration *via* the 4,6-orthoacetate to yield the 2,3,3′,4′,6-pentaacetate (Scheme 2.2).

The most important reaction which leads to sweeter derivatives is the simple substitution of 1′-OH by chlorine. Compounds (VII) and (XV) are hydrolysed neither by α-galactosidase nor by β-fructofuranosidase and are more stable towards acid hydrolysis than the parent sucrose [3, 41]. Favourable properties, especially the low toxicity, predestined the trichloro derivative (XV) for the role of non-carious, non-nutritious and very potent sweetener. The derivatization of (XV) at C_3 and C_4, yielded a whole new group of promising compounds.

The reaction of 6-acetylsucrose with sulphuryl chloride at low temperature in a mixture of pyridine and chloroform, followed by a deacetylation, furnished two products, 4-chloro-4-deoxy-α-D-galactopyranosyl-1,4,6-trichloro-1,4,6-trideoxy-β-D-sorbofuranoside (XVI) as well as 4-chloro-4-deoxy-α-D-galactopyranosyl-1,4,6-trichloro-

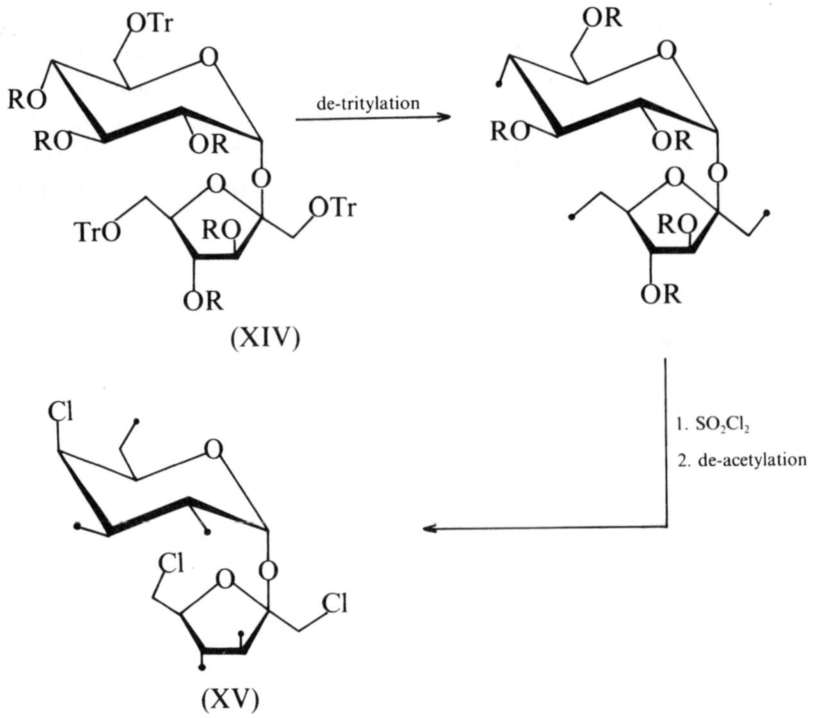

OTr
RO
RO
OR
O
O
TrO
RO
OTr
OR
(XIV)
de-tritylation
OR
RO
OR
O
O
RO
OR
1. SO_2Cl_2
2. de-acetylation
Cl
O
Cl
O
O
Cl
(XV)

R = CH_3CO

Scheme 2.2

(XXI)
(XXII)
Cl
O
O
Cl
O
Cl
Cl
(XVI)
Cl
O
O
Cl
O
Cl
Cl
(XVII)
= OH

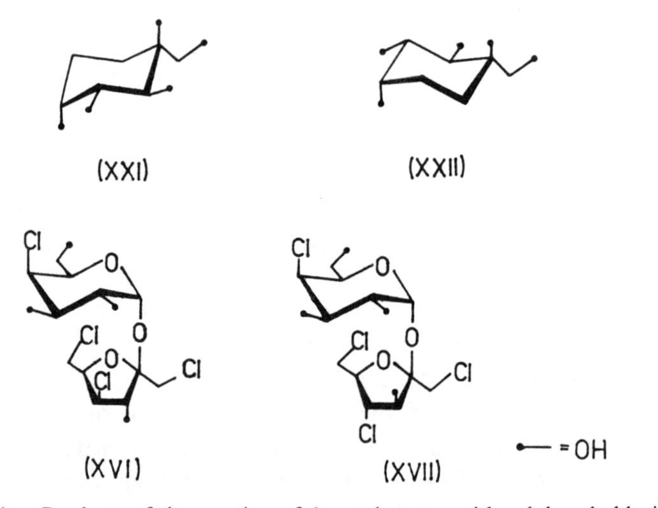

Fig. 2.6 — Products of the reaction of 6-acetylsucrose with sulphuryl chloride.

1,4,6-trideoxy-β-D-fructofuranoside (XVII) [32] (Fig. 2.6). Both tetrachloro derivatives are intensely sweet but their relative sweetness is an order of magnitude apart due to the configurational difference (rel. sweetness XVI — 200, XVII — 2200).

The introduction of chlorine at C_4, during the synthesis of (XVI) and (XVII), required an opening of the intermediate 3,4-epoxide by the chloride anion — a route that was previously suggested in the literature [20, 47].

Bromo substituted sucroses such as 1′,4,6′-tribomo-1′4,6′-trideoxy-galactosucrose can be as sweet as the better known chlorinated sucroses [48].

2.1.1.2 Spectral studies of sucrose

^{1}H NMR spectroscopy has currently assumed a central position in the spectral analysis of sucrose. It can identify the substitution pattern as well as the stereochemistry at any carbon atom and is well suited for the assessment of conformations and conformational equilibria [49, 50]. As an example, let us examine the proton resonance spectrum of sucrose octaacetate [51, 52]. The $^4C_1$ conformation of the α-D-glucopyranose unit can be inferred from the first order coupling constants ($J_{1,2}$ = 3.7 Hz; $J_{2,3}$ = 9.5 Hz; $J_{4,5}$ = 9.7 Hz). Similarly, the coupling constants of the β-D-fructofuranose unit ($J_{3',4'}$ = 5.5 Hz, $J_{4',5'}$ = 5.1 Hz) indicate the presence of a conformation that has atoms C_2, C_3, C_5, O_5 in a plane, whilst the C_4 is out of the plane.

^{13}C NMR is another valuable analytical tool that furnishes information on the position of substituents and the stereochemistry [53, 54].

The effect of substituents on the ^{13}C NMR spectra was examined by several authors [55–58]. Ball *et al.* [58] observed a downfield shift of a carbon atom that carries a tosylate by as much as 5 ppm whilst the C atom in β-position shifts upfield by 2–3 ppm. These findings enabled the identification of a by-product of mesylation, 2,6,6′-tri-*O*-methyl-sulphonyl sucrose. Similar empirical rules have been always very useful in the interpretation of spectra and may, in the future, help to solve structures of complicated sucrose derivatives.

Mass spectrometry is also counted as one of the routine methods of analysis of some saccharide derivatives. In general, mass spectra of sucrose seldom display the parent molecular ion because of the presence of a very weak glycosic bond. The two subunits, glycosyl cations, are well represented in the fragmentation pattern and correspond to the hexapyranosyl (Gl^+) and ketofuranosyl (Fr^+) residues. Both fragment further by individual pathways [59, 60]. Kochetkov and his co-workers [59] were the first to suggest a fragmentation scheme for disaccharides.

An advantage of mass spectrometry manifests itself specially in cases in which, as in a case of sucrose, the disaccharide cleaves into two primary ions of different stability. The fact that Fr^+ is more stable, and therefore more intense, than Gl^+ [61] indicates much better charge delocalization at the tertiary carbon atom [37].

2.1.1.3 Sucrose in human diet

The energy content of sucrose is 16 kJ/g [62]. In spite of this, the human organism covers only a small part of its total energy demand (approximately 500 g of saccharides per day) by metabolizing directly sucrose — the rest is provided by starch from various sources such as potato and flour [63].

The metabolism of sucrose and starch have not yet been clearly understood and views on the mechanism differ. An early, but still accepted hypothesis contends that starch can be utilized by human organism only after it has been broken up into simple saccharides. Sucrose and other simple saccharides metabolize faster. This, combined with favourable taste, has made sucrose an important component of human diet. However, a slow metabolism of starch can sustain certain levels of blood sugar for much longer periods.

Unfortunately, fast metabolism of sucrose not only gives a fast energy release but is also responsible for the unwanted conversion of excess blood sugar into fats [63–66]. It follows that in order to enjoy a balanced energy input, only 25 % of required saccharides should be supplied in the form of sucrose.

The relationship between sucrose intake and the formation of fats is governed by several factors, such as the character and quantity (saturated or unsaturated) of consumed fat, age and sex, period of observation, total energy intake, as well as the original reference level of the triacetyl glycerols [63, 67–69].

Many studies that led to the conclusion that simple saccharides were digested and absorbed faster than polysaccharides failed to take into account all relevant circumstances. For instance, in some cases uncooked starch elicited lower glycemic response [70] because it took longer to hydrolyse and absorb it [71]. On the other hand, cooked starch-containing foodstuffs (potatoes, rice) showed similar or only marginally slower glycemic response than glucose [70, 72]. Indeed, the glycemic response of cooked starch equalled or was better than that of sucrose [70, 72, 73]. The role of sucrose in human diet has remained in the focus of interest especially after alternative sweeteners (developed originally for diabetics) became widely available and their long-term effects appreciated. The experiments with NIDD — non-insulin-dependent diabetics — have shown that modest amounts of sucrose (up to 28 g — 7 teaspoonfulls) were an acceptable daily dose [74]. However, the recommended daily dose for healthy humans should not exceed 75 g, that is 28 kg in a year.

The role of sucrose in the human diet became a hotly debated topic by the international scientific community and food industrialists. Advances in the knowledge of the biosynthesis of sucrose have helped us to understand the mechanism of sucrose resorption in human organism. Sucrose metabolism has also been studied in relation to the vitamin balance in the organism, the occurrence of dental caries, optimal ratio of vitamins, especially those of group B, phosphorus, calcium and other food components, in order to ascertain the conditions in which saccharides are utilized best.

Producers of confectioneries and canned food industry consume huge quantities of sugar. Stringent requirements of the bacteriological purity of sugar are demanded by the latter producers. Although a 30% solution of sugar resists fermentation and 60% solution is stable even in the presence of bacteria, yeast strains *Zygosaccharomyces* are still active even in a solution with sugar concentration higher than 70%, e.g. honey with a little moisture. Thermophilic bacteria thrive even on crystalline sugar.

2.1.1.4 Application forms

(1) Raw or cooked sugar beets are used in minimal quantities in several types of sweet and sour salads. Unpleasant taste and fibrous texture handicap its more widespread use.

Completely or partially dried beet flour is added to biscuits, gingerbread and bread; it also finds some use in beer brewing.

(2) The raw beet juice has been manufactured industrially or at home by extraction with boiling water. It must be concentrated before it can be used. Although it is an interesting intermediate in its own right, larger production of raw juice would be too costly because the beets would have to be sorted in order to remove dirty and rotten starting material.

(3) Crude unrefined sugar has often been hailed as being more natural and healthier than the refined sugar. Some prefer it in their coffee or biscuits because of its flavour. The brown colour of raw sugar is due to organic colouring matters — for instance caramel — which have no dietary value. Apart from being contaminated by both organic and inorganic impurities (cane trash, soil, other dirt, bacteria, moulds (fungi) and live sugar lice), raw sugar is impractical to keep and distribute because it hardens when it dries. It would have

to be sold either in small lumps or as cubes, both forms incurring an extra cost.

(4) Refined and chemically pure sugar contains no less than 99.8 % of sucrose, the rest are inorganic impurities (ash), traces of organic impurities and water. The composition of all its commerical forms (loaves, cubes, crystals, granules) is practically the same. The final product is sometimes dyed ultramarine as it looks more attractive in some applications such as confectionery, sparkling wines, children's sweets and canned fruits.

(5) The dietary value of pure sucrose can be enhanced by addition of other dietetically beneficial matter such as calcium and magnesium salts, sodium chloride, organic acids and vitamins. However, the mixture is often too expensive to produce on any larger scale, even if it is possibly beneficial to health.

(6) Consumable sweet syrups and molasses are produced from syrups and white refined sugar. They have a pleasant taste and smell and are widely used in the production of pastries and canned fruits.

(7) A speciality product — liquid sugar — is sold in very large volumes predominantly to industry. It is approximately 70 % aqueous solution of sucrose and invert sugar (glucose and fructose).

(8) Instant sugar, another speciality product, that results when pulverized sugar is damped or sprayed with sugar solution and the resulting agglomerates are then dried, crushed and sorted. Because of its high porosity, the volume of instant sugar is about twice as large as that of normal sugar. It has improved solubility, it is not hygroscopic and is stable to temperature changes.

2.1.1.5 Principles of the technology of sugar production

Although many plants contain quite a lot of sugar, the commercial production in large quantities is possible only at reasonable cost from sugar cane and sugar beet. Sugar cane is grown in hot countries whilst sugar beet only in Temperate Zone countries (Scheme 2.3).

The technological process of sugar extraction from the beet [75] involves the following steps: extraction of sugar from beet cells by water, purification of the dilute juice, carbonation of the raw juice by the addition of slaked lime and gaseous carbon dioxide and then followed by filtration so that the chalk with the entangled impurities is removed. The clear juice is then concentrated, boiled in vacuum pans, and centrifuged to separate sugar from molasses. The raw sugar that is obtained is finally purified.

The sugar beet root contains about 75 % water and 25 % dry matter. The latter contains about 15 to 17 % sucrose, the rest are various

Technology of sucrose production from sugar beet

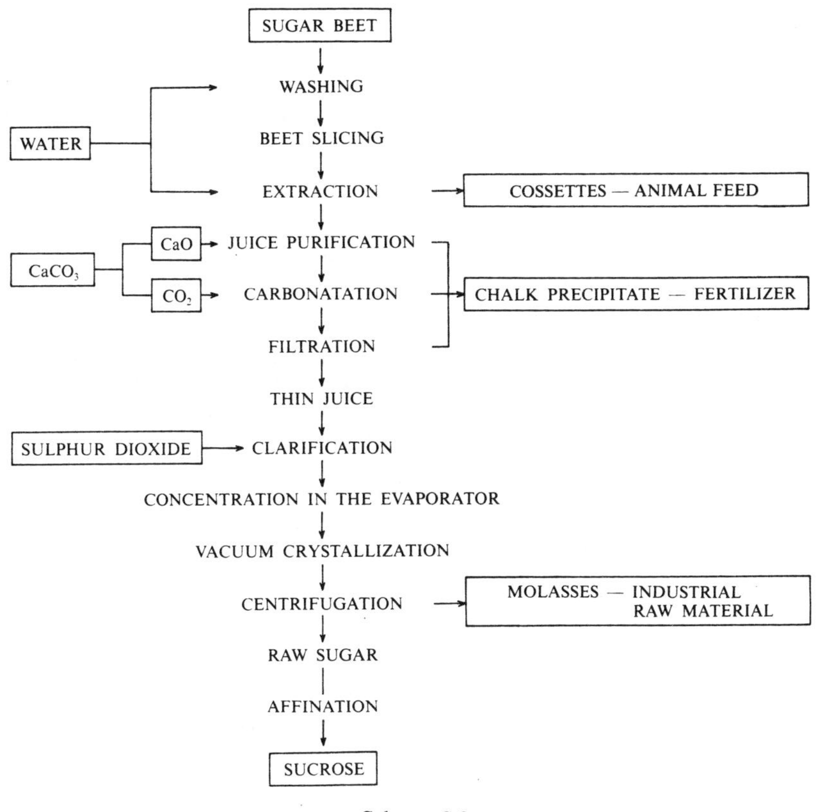

Scheme 2.3

components of inorganic and organic origin. Most of them are removed in the course of fining.

The beet is weighed and sampled on arrival at the factory and then tipped into large concrete silos which have a narrow channel of running water on the bottom. Water not only transports the beet, it also washes mud and stones from the beet roots. Beet passes through plant combs and beet washers onto the picking tables where stones, bits of wood and other unwanted material is removed by hand. It continues to the slicing machines in which it is chopped into shreds or cossettes.

Sliced beet is fed into a battery of connected extraction tanks (10–16) where the dilute juice is separated from the extracted shreds or pulp.

The beet juice is a dark grey to brown liquid which contains about 13 to 15 % of sugar, 3 % of impurities and the rest is water. The juice is then treated with slaked lime (produced on site in a lime kiln) and CO_2 gas in a process called carbonation. The gas combines with calcium hydroxide to form a fine precipitate of chalk which encapsulates most of the gummy matter as well as some of the colour. Subsequent filtration removes the chalk and impurities, yielding a light-coloured filtrate which contains about 15 % of sugar. Further decolourization is achieved by gaseous sulphur dioxide. The clear thin juice is then fed into a multistep evaporator where it is concentrated into a dark red, heavy and thick juice. Another decolourization is carried out either by SO_2, charcoal or ion exchangers, depending on the colour intensity.

Heavy juice is then put into vacuum pans where sugar crystallizes out. The mixture of crystals and syrup which comes out of the pans, called massecuite, is fed into centrifuges and yields the first crop of sugar. The separated syrup is returned to the vacuum pans for concentration to give a second crop. White crystals from the first crop are steam washed, cooled and after the removal of fine dust, sorted and packed. The second crop of crystals is still covered with some syrup (wet sugar), the second crop of syrup is greenish and is recycled (mixed with raw syrup) to give the third crop of sugar and the dark exhausted syrup — molasses. The third crop sugar crystal is almost always evenly and well formed but too small to centrifuge satisfactorily. Therefore, the syrup is generally seeded with crystals from earlier crops.

Molasses is a complicated solution of salts, organic matter, sucrose and invert sugar in water. Although the molasses in its 'exhausted condition' still contains about 50 % of total sucrose, the syrup becomes so thick and viscous that crystallization virtually stops. However, such a solution is a valuable feedstock for other branches of food industry — it is used for production of alcohol, yeast, citric and lactic acids, and it is also added to cattle fodder.

The exhausted pulp from the first step which still contains a small amount of sugar (0.3–0.5 %) is compressed to remove the excess water and then sold wet as cattle fodder, or dried and stored and then sold as cattle food (often mixed with molasses). The scum that separates from the slaked lime by skimming or by sedimentation makes good calcium fertilizer; when chlorinated, it gives the so-called saturation VK chalk. The latter is used, most appropriately, for the pest protection of beets during storage.

The refined sugar, after the removal of fine dust, must be separated into fine and coarse grain sugar and then stored in sugar silos. When sugar is dampened, compressed into forms under pressure and then dried, it retains its shape. In order to ascertain that the cubes never have any tinge of yellow colour, sugar has to be boiled down from specially good syrup. The cubes, although not very hard, are packed in tight stacks in the packets so that they cannot move and thus travel very well.

Any inevitable breakage during cube production is removed and converted by milling into a variety of high-quality speciality products, such as icing sugar and fine caster, or pulverized sugar which is mixed with about 1 to 3 % of starch (or $Ca_3(PO_4)_2$) in order to improve its keeping qualities. Other products are crystal sugar (of various crystal size), sugar loaves, Golden syrup, liquid sugar, Nibs and yellow sugar.

2.1.1.6 Projections for the production and consumption of sucrose

The main competitors to sucrose, as the main sweetener in the food industry, are the starch produced sweeteners, especially starch syrup which has a higher fructose content and thus is used for the production of food with reduced energy content and food for diabetics.

When looking for alternatives to sucrose, one must not forget other qualities that sucrose contributes to the overall appearance and quality of the product, such as texture, added weight and preserving properties. Thus when sucrose is replaced by a synthetic sweetener, an immediate need arises for extra stabilizers, fillers and gelation agents in order to simulate the overall quality that is lost by the replacement. Industrialized countries utilize a part of their production of sucrose for the manufacture of other chemicals (e.g. surfactants). This utilization receives an additional impetus from the changing situation in the global oil market because of the gradual exhaustion of this non-renewable commodity. This makes sucrose, a renewable commodity, very competitive in many other applications as well.

The use of sucrose in Czechoslovakia is also diverted [76] to other applications, such as the production of oxalic acid and glucose, which is, in turn, fermented to lysin. A small proportion of sucrose is converted into antibiotics.

2.1.2 D-Glucose (dextrose) — the most important aldohexose

D-Glucopyranose forms a part of many natural compounds. It can be produced most easily from starch by acid hydrolysis with dilute hy-

Fig. 2. 7 — Anomers of D-glucopyranose.

drochloric acid. It is easily separated from the second product of hydrolysis, D-fructose, which crystallizes with great difficulty. Both anomers can be obtained, the α-anomer (XVIII) (Fig. 2.7) by crystallization from hot ethanol [77, 78] or from water by seeding [79] and the β-anomer by crystallization from hot water or by precipitating from dilute solutions in acetic acid or pyridine by addition of ethanol. Both anomers form colourless crystals that melt at 146 °C (anhydrous α-anomer [80]) and 148–155 °C (β-anomer). The α-anomer also crystallizes as a monohydrate which melts at 83 °C [81].

The major usage of glucose is in dietetics. It is an excellent energy source both for children and adults because of its very fast uptake by the human organism. It is used in remedies for tiredness, in soft drinks, chocolate and in beer. Glucose is also present in many medicinal applications and can be introduced into the body either by injection or by infusion. It is also found in many ointments.

2.1.3 D-Fructose (laevulose)

D-Fructose (laevulose or fruit sugar) is the most important ketohexose (Table 2.3). It is found mainly in honey and fruit juices (Table 2.4); it is also available from the hydrolysis of sucrose and inuline (Table 2.5) [82].

Table 2.3 — Physical properties of some monosaccharides, disaccharides and polyols

Compound	Formula	Mol. wt.	M.p. (°C)
D-Glucose	$C_6H_{12}O_6$	180.1	153–156
L-Glucose	$C_6H_{12}O_6$	180.1	153–154
D-Fructose	$C_6H_{12}O_6$	180.1	59
D-Mannose	$C_6H_{12}O_6$	180.1	129–132
D-Sorbose	$C_6H_{12}O_6$	180.1	162–164
D-Galactose	$C_6H_{12}O_6$	180.1	166
D-Xylose	$C_5H_{10}O_5$	150.1	156–158
Saccharose	$C_{12}H_{22}O_{11}$	342.3	160
Lactose	$C_{12}H_{22}O_{11}$	342.3	203
Maltose	$C_{12}H_{22}O_{11}$	342.3	130
D-Glucitol	$C_6H_{14}O_6$	182.1	98–100
D-Mannitol	$C_6H_{24}O_6$	182.1	167–170
Xylitol	$C_5H_{12}O_5$	152.1	92–95
Lactitol . H_2O	$C_{12}H_{26}O_{12}$	362.3	115–125
Maltitol	$C_{12}H_{24}O_{11}$	344.3	149–152

Table 2.4 — Composition of honey

Compound	Content (%)
D-Fructose	40
D-Glucose	35
Water	18
Other saccharides	4
Other compounds	3

Table 2.5 — Occurrence of D-fructose in fruit

Fruit	Content in fruit (%)	Content in dry matter (%)
Apples	6.04	37.8
Pears	6.77	49.9
Blackberries	2.15	14.1
Sweet cherries	7.38	32.9
Strawberries	2.40	25.4
Raspberries	4.84	17.2
Gooseberries	3.90	26.3
Currants	3.68	20.8
Grapefruit	7.84	41.0

2.1.3.1 Sweet taste of fructose and its analogues

D-Fructose (Fig. 2.8) is the sweetest natural saccharide, its intense sweetness originates from the β-D-fructopyranose unit [83–86]. The anomeric HO-2 and the CH_2OH group in the β-D-fructopyranose have been designated as the AH–B system which is responsible for the sweetness [83, 87–89]. The principle of the stability of rotational isomers supports this assignment also [90]. The third binding point (γ) has been localized at the C_6 atom, thus completing the three-point sweet unit (XX) [91] (Fig. 2.9).

Many papers have dealt with the problem of D-fructose structure [92] and its relationship to sweetness [93–95]. Another problem that continues to fascinate the sugar and food chemists relates to pseudo-

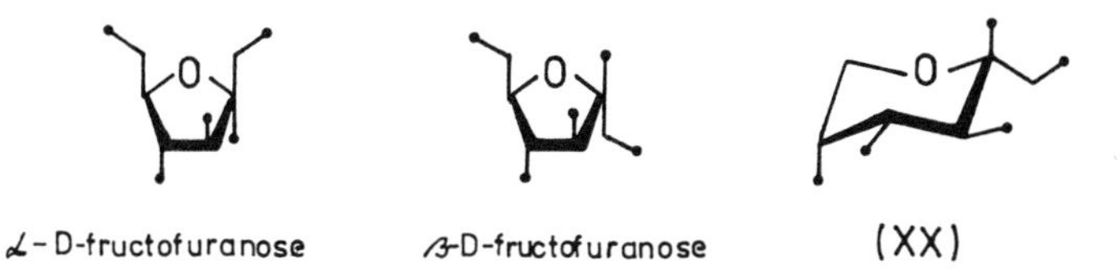

Fig. 2.8 — Anomers of D-fructose.

saccharides, particularly the as yet unanswered question of the organoleptic properties of D- and L-fructoses. Experiments revealed that pseudo-β-L,D-fructopyranose [96–98] as well as the synthetic enantio-

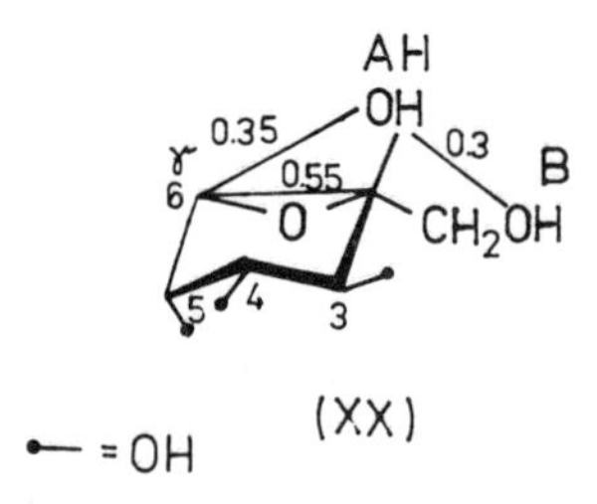

Fig. 2.9 — AH–B–γ system of β-D-fructopyranose.

mers [98], pseudo-β-D- (XXI) and L-fructopyranose (XXII) (Fig. 2.10), are about equally as sweet as fructose. β-D-Fructopyranose exists in the $^2C_5$ conformation only in the crystal, rapid mutarotation takes place on dissolving in water and results in the formation of three, or even four, fructose tautomers [83]. The extent of mutarotation can be determined by measuring the optical rotation of a solution of pure fructose in water. A more sophisticated approach that utilizes gas-liquid chromatography and nuclear magnetic resonance has been employed to determine the tautomeric equilibrium of a 20 % solution of fructose in D_2O at 22 °C; it contains 73 % β-D-fructopyranose, 6 % α-D-fructofuranose and 21 % β-D-fructofuranose.

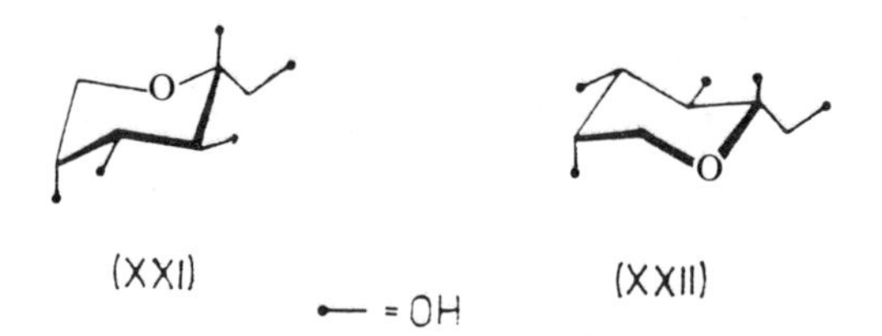

Fig. 2.10 — Pseudofructopyranoses.

The temperature (Table 2.6), pH and the concentration of the solution are the most significant parameters that influence the magnitude of the perceived sweetness of a freshly prepared aqueous solution of fructose. However, only the temperature exerts a major effect on the sweetness because it affects the mutarotational equilibrium [99].

It follows from the above that the maximal sweet potency of fructose can be derived by consuming crystalline pure fructose in order to reduce

Table 2.6 — Composition of equilibrated solutions of D-fructose at different temperatures

Temperature (°C)	α-D-Fructofuranose (%)	β-D-Fructofuranose (%)	α-D-Fructopyranose (%)
20	7	24	69
40	7	31	62
60	9	33	58
80	11	38	51

the dissolution-initiated mutarotation. The loss of sweetness, due to the shift in the position of the equilibrium towards furanose forms, can be minimized by cooling the solutions and by keeping them slightly acidic. Thus practical experience shows that lemon flavoured soft drink mixtures that are sweetened by pure crystalline fructose and contain the usual amount of the sour component can preserve the required sweetness even at 50 % reduction of their energy content. In contrast, the least effective use of fructose as a sweetener is in hot drinks such as coffee and tea, when the large extent of mutarotation reduces its sweetness relative to that of sucrose. Another way to utilize the sweet properties of fructose when used in baking is to consume the pastry only when it has cooled down sufficiently.

2.1.3.2 D-Fructose in human diet

The utilization of D-fructose in human beings starts in the gastrointestinal tract from which it is absorbed into the intestinal epithelial cells. The exact mechanism that is involved is not certain [99]. After the absorption, which takes place at a slower rate than that for sucrose or glucose, it is rapidly metabolized in the liver. Significantly, its entry into the liver and subsequent phosphorylation is insulin-independent [100]. When comparing D-glucose, sucrose and D-fructose, Bohannon *et al.* [101] came to the conclusion that fructose ingestion causes less marked baseline variations in the blood glucose, insulin, glucogen and growth hormone levels in normal humans than does sucrose or glucose. Most diabetics tolerate fructose quite well [100]. Retarded fructose adsorption in sensitive individuals may result in osmotic diarrhoea, especially when only fructose is consumed. When accompanied by D-glucose [102] as in fruit, or by isoglucose, it is adsorbed much better.

The increased awareness of the harmful side-effects of diabetes and the necessity for keeping the blood sugar levels of diabetics as close to normal without inducing severe hypoglycemia, has led diabetologists and food technologists to renew investigations into the use of fructose as a preferred sweetener for diabetics [99].

Sorbitol has been widely accepted as a sugar substitute for the use in a well managed diabetic diet [99]. After the absorption, it is rapidly converted into fructose by the enzymatic action of sorbitol dehydrogenase; thereafter, its metabolism is identical with that of the ingested fructose. The important difference is that the perception of sweetness of sorbitol is only about one third of that given by an equal amount of fructose.

Pure crystalline fructose can be used effectively as a sweetener in any food. Its water solubility is excellent and its ability to absorb moisture can be utilized in some applications. The solutions of fructose are less viscous than those of sucrose in identical concentrations. Furthermore, fructose amplifies the fruity flavour and taste.

Some of the commercial applications of fructose are: dietetic cake mixes, desserts, puddings, gelatins, dietetic and regular sweets, gums, tabletop sweeteners, frozen desserts, diet soft drinks, dietary meal replacements and powdered beverages. Fructose containing foodstuffs are designated as being low-calorie; their energy content is one third less in comparison when sucrose is used as the sweetening agent. A real advance in producing a low-calorie and pleasantly tasting sweetener has been achieved by mixing fructose, the sweetest natural sugar, with artificial sweeteners such as saccharin [103] or aspartame [99] and others. Synergism was observed in some cases.

2.1.3.3 Manufacture of fructose

Inulin, the fructose precursor, can be obtained from carbohydrates of various plants such as the bulbs of dahlias, Jerusalem artichokes and Hawaian Ti plant. Controlled hydrolysis of the polyfructan inulin cleaves the β-(2→1)glycosidic bonds yielding mainly fructose which is converted to the more stable fructopyranose anomer. However, this process never reached industrial utilization.

Serious efforts to produce pure fructose in the United States were taken in the early 1950s when Holstein and Holsing patented their method [104]. It involves the inversion of sucrose either enzymatically or by acid hydrolysis into invert sugar which is then oxidized enzymatically to gluconic acid which is crystallized out as the sodium salt, and crystalline fructose of high purity is obtained from mother liquor.

Industrial production of D-fructose started in Europe in 1960s. The process was based on the hydrolysis of sucrose into monosaccharides which were then separated and purified by ion-exclusion techniques.

Fructose was obtained by carefully controlled crystallization; the processing time exceeded one week.

The world's largest plant for the production of pure crystalline fructose came on stream in 1981 in Thompson, Illinois on the Mississippi River. As the plant is located in the corn belt, it utilizes liquid dextroses as the raw starting material. After purification and enzymatic isomerization, the traditional techniques are used in order to produce very high-quality crystalline fructose. The manufacturing time was reduced to about five days [99].

An alternative method that is based on an extraction of fructose from topinambur *(Helianthus tuberosus)* was suggested by Barta [105]. Crystalline D-fructose is obtained by crystallization from aqueous solutions.

2.1.3.4 Production and consumption prospects

The current worldwide consumption of pure crystalline fructose does not exceed 30,000 tonnes/year. The largest consumer is the United States which supplies their needs from the domestic production. The main sources of supply prior to 1980 were Finland and Germany — the International Fructose Association has seven members.

It is likely that the demand for fructose will continue to increase as the emphasis for nutritionally balanced foods gains wider acceptance. However, the production of crystalline fructose is very capital intensive and requires extraordinary amounts of increasingly expensive energy. Thus it is most unlikely that its cost will decrease dramatically in the near future.

It should be realized that fructose is not a sweetener for all occasions and all products. However, fructose can be included as a unique ingredient which, because of its presence, will make the product more valuable [99].

2.1.4 Starch hydrolysates (starch syrups)

Limited starch hydrolysis occupies an important place amongst the many hydrolytic procedures of higher polysaccharides. Its product is called *hydrogenated starch hydrolysates* (HSH) which is a mixture of mono-, di- and polysaccharides. Glucose and fructose syrups are the two best known HSH products and their importance on the speciality sweetener market has steadily increased. These syrups were originally manufactured from potato starch — present technology utilizes corn (maize) starch. The hydrolysis of starch is rather a complicated process which is promoted by acids, enzymes or by both in combination. Starch is first hydrolysed into a syrup consisting of glucose, maltose and higher saccharides. The continuous hydrolytic process can be interrupted at various stages which yield hydrolysates with different dextrose equivalents (DE). Ultimately the final product of hydrolysis is D-glucose — dextrose which is the basic building block of starch. Table 2.7 classifies the hydrolytic products according to their DE from which it is possible to calculate the content of reducing sugars in the dry matter (Table 2.8).

Table 2.7 — Relative abundancy of saccharides in various starch syrups

	Low degree of conversion	High degree of conversion	Maltose syrup	Very high degree of conversion	Fructose syrup
Glucose (%)	16.2	43.3	9.6	92.3	50.4
Fructose (%)	—	—	—	—	42.6
Maltose (%)	10.6	19.6	40.1	3.6	4.1
DP 3%	1.4	7.8	15.5	0.8	0.9
DP 4%	6.4	3.7	6.8	0.7	0.2
DP 5%	6.9	2.6	1.8	0.8	0.2
Higher saccharides	48.5	13.0	26.2	1.8	1.6

DP — Degree of polymerization.

Table 2.8 — Composition of starch syrup with dextrose equivalent DE* 92

Dry matter	7.4 ± 0.5
Monosaccharides, relative to dry matter (%)	min 92
Ash (sulphates) (%)	max 0.05
Heavy metals (mg/kg)	max 2.6
SO_2 (mg/kg)	3–4
pH (non-diluted syrup)	3.5–4.5

* Degree of saccharification (DE = dextrose equivalent).

Other products that are accessible from starch by a combination of hydrolysis and hydrogenation represent a whole group of speciality sweeteners [106] (Scheme 2.4).

One prominent product in this group of sweeteners is fructose syrup which has revolutionized the sweetener industry in the United States by providing a viable and economical alternative to sucrose, both in regular and reduced-calorie products. The advances in enzyme technology that took place in the late 1960s enabled the conversion of corn starch into syrups that are high in fructose content. These high-fructose

Alternative sweeteners manufactured from starch

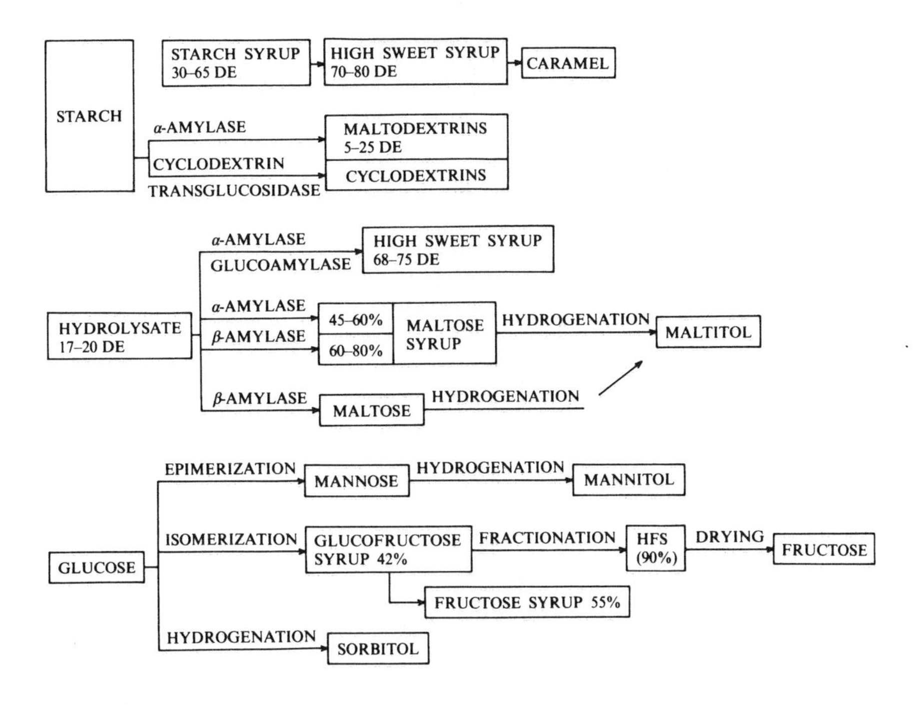

Scheme 2.4

corn syrups (HFCS) are, in fact, the result of an enzyme isomerization of dextrose (D-glucose) syrups (Scheme 2.5).

The isomerized product resembles invert sugar in as much that 70–72 % of its dry matter is D-fructose, a sufficient amount to possess sucrose-like sweetness. Fructose syrup has been marketed under various brand names such as Isomerose, Isosweet, Corn Sweet, Meliose, Ameriose, More-Sweet, and is often designated as HFCS or HFS (high-fructose syrup). It has been designated in German literature also as Isosirup and Isoglukose [107]. The development of fructose syrup technology which utilizes the immobilized enzymes is considered as one of the most significant successes that have been achieved in the last two decades in the food industry. The incentive to use HFCS is controlled overwhelmingly by economic considerations and it varies with the supply/demand relationship. HFCS prices are generally 15–30 % lower than those of sugar. The market share of HFCS is still growing, from 22 % in 1982 to 30 % in 1987.

Enzyme process for high dextrose syrup

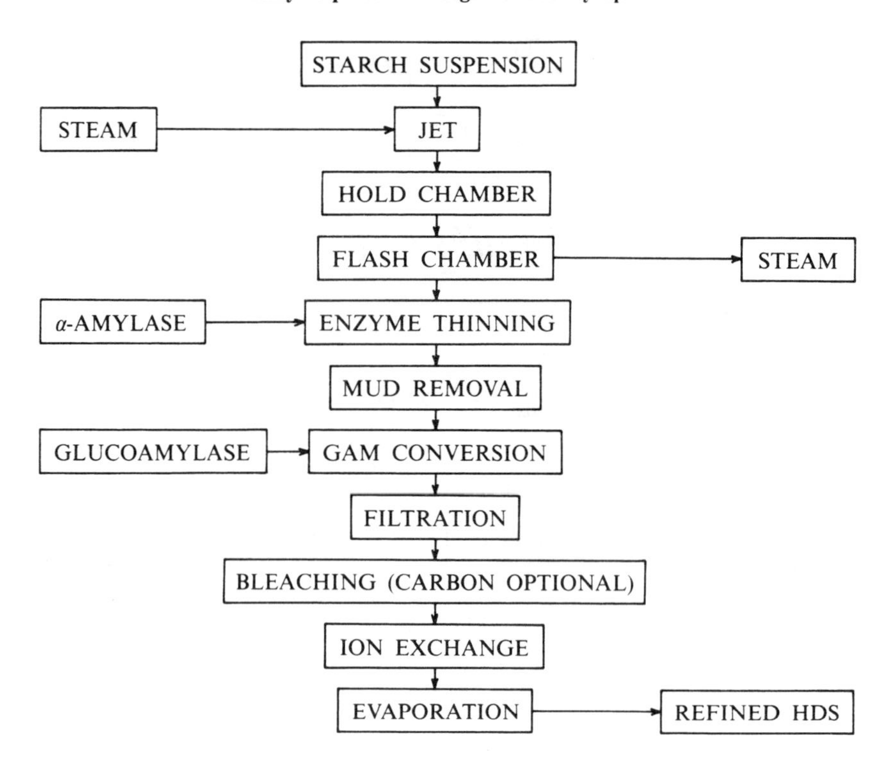

Scheme 2.5

2.1.4.1 Properties and utilization of fructose syrup

The 'first generation' HFCS contained 42 % fructose and 50 % dextrose and was produced on a commercial basis in the early 1970s. It found widespread use in baked and other food products. The 42 % HFCS served only as a partial replacement for sucrose in still and carbonated beverages. It was not until 'second generation' HFCS (composed of 55 % fructose and 40 % dextrose) appeared in the mid-1970s that its use in the soft drink industry became more widespread. The main reason for this was the fact that the composition and the taste of the new HFCS was very close to that of the totally hydrolysed medium invert syrup which was an established product in the soft drink industry.

The second generation HFCS have already found wide acceptance on the US market for the low-calorie products, called light food. They are not strictly dietetic products but they have lower energy content than normal food and drink because of the change in composition.

The industry also produces a 90 % fructose syrup which is sweeter than sucrose. It is used mainly in calorie-reduced products (table syrup, dairy products, canned fruits, salad dressings and other) which contain up to 50 % less calories [108].

Increased sweetness is not the only asset of HFCS sweeteners. The composition of HFCS remains constant in contrast to sucrose or invert sugar sweetened drinks, which can change during storage.

The 90 % HFCS also allows baked goods to hold their moisture, it controls the hardness and ice crystallization in frozen desserts and the texture in candy bars.

2.1.4.2 Technology of fructose syrup

Both dextrose and fructose syrups derive from starch, a polysaccharide of plant origin. However, not all starch-containing plants qualify as the starting material; the best known sources with high starch content, availability and low price are corn, wheat, potatoes, rice and manioc. The choice between these starting materials is made after taking into account the availability, the quantity required, production costs and the specific properties of the produced starch which are pertinent to its subsequent saccharification. Thus in Europe, starch is at present produced from corn.

The technology of fructose syrups relies on acid (HCl) or enzyme catalysed hydrolysis of starch. In the enzyme process, amylase converts starch first into a product mixture with DE value of 18–22; the products of amyloglucosidase hydrolysis have higher dextrose content (DE 97–98). A typical HDS (high-dextrose syrup) process is shown in Scheme 2.5. The starch slurry is thinned with α-amylase *(Bacillus subtilis)* which can be added either at the beginning, or as shown in Scheme 2.5. Still further hydrolysis can be achieved by the addition of the dextrose producing glucoamylase. The refining process consists of filtration, carbon treatment followed by ion exchange and concentration by evaporation.

The production of fructose syrup need not start from starch. Various processes exist that can convert starch-containing materials directly into fructose syrup, i.e. without the isolation of the starch itself. The reason why these procedures have never been commercially utilized is the high cost of purification of the direct hydrolysis syrups [109–111].

The hydrolytic product obtained in the first step of the HDS process serves as the starting material for fructose syrups. The process involves a conversion of a part of the D-glucose present in HDS into D-fructose. The commercialization of this process became viable after the introduction of immobilized enzymes into industrial production. Glucose is isomerized by a passage through a bed of immobilized isomerase (glucoisomerase) which is produced by various microorganisms (*Lactobacillus, Pseudomonas, Pasteurella, Leuconostol, Streptomyces, Aerobacter* and others). These are, as a rule, intracellular enzymes which vary in their properties such as growth medium demands, dependence on metal ions, pH and temperature. The entire process of enzymatic hydrolysis lends itself to a high degree of automation and on-line process control [112].

The technology of enzyme immobilization for the production of HFCS has been patented [113, 114]. The American fructose syrup producers operate their own facilities for the glucoisomerase production. A flowchart of the isomerization and fractionation procedures is depicted in Scheme 2.6.

A significant progress has been made over the last few years in the improvement of quality and consistency of thc product. The lowering of the concentration of acetaldehyde represents only one indicator of

Production of HFCS

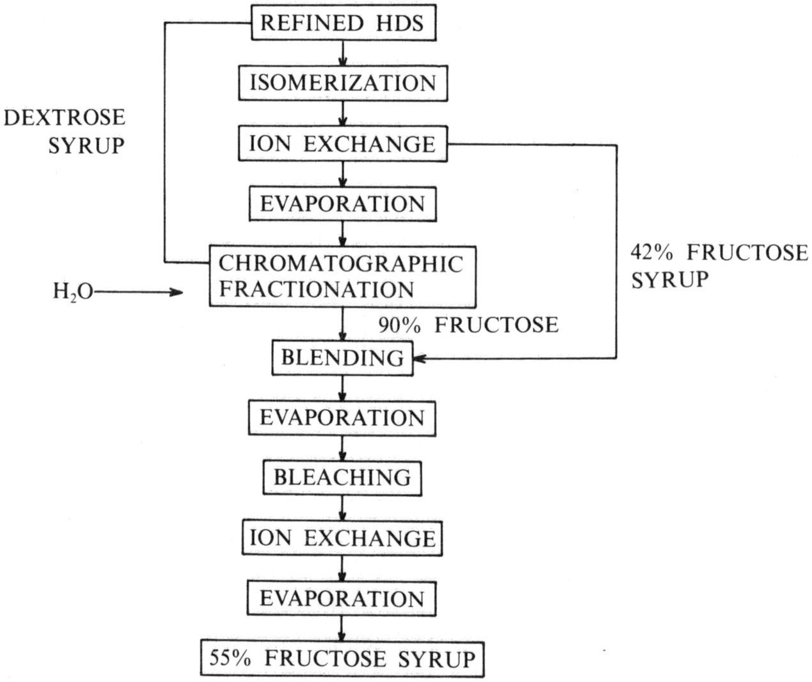

REFINED HDS
ISOMERIZATION
DEXTROSE SYRUP
ION EXCHANGE
EVAPORATION
42% FRUCTOSE SYRUP
H_2O → CHROMATOGRAPHIC FRACTIONATION
90% FRUCTOSE
BLENDING
EVAPORATION
BLEACHING
ION EXCHANGE
EVAPORATION
55% FRUCTOSE SYRUP

Scheme 2.6

these changes. Acetaldehyde, which is formed in the HFCS production process, affects the flavour of cola beverages even at very low concentration levels [115].

2.1.4.3 Production and consumption prospects of HFCS

The most important challenge that the HFCS industry will have to face in the next decade is the increase in the efficiency of the process and to produce dry, pure crystalline fructose. Advanced refining and fractionation techniques are being developed, as well as new enzyme systems which can produce higher fructose yields at lower cost. Improved separation techniques could produce as much as 98 % yields.

Owing to the large progress achieved over the past decade, fructose syrups became a major sweetener in the United States. It functions in the same way as sucrose in many applications and it will most likely continue to replace sucrose, specially in the soft drink industry. This is true not only for the US but for Canada as well. On the other hand, the facilities that were built in Europe have been impeded in their development by restrictive governmental policies.

As the HFCS technology is bound to show still higher yields and a better quality product, the consumption of fructose syrups is expected to increase and ultimately, fructose in dry form made by HFCS processes will become competitive with sucrose.

2.1.5 Malt extract and malt syrup

Aqueous malt extract contains a variable composition of water-soluble matter, produced by the action of malt enzymes, which consists of various saccharides, proteins, minerals and enzymes. The dry matter contains 4–5 % sucrose, in addition to fructose, maltose and glucose.

Due to the presence of enzymes, malt extracts facilitate the degradation of starch, and thus also its digestibility. Their main application area is in the manufacture of sweets; it is also used as a nutrient for yeast cultures in baking technology, as well as in the preparation of special foods for young children and babies.

Maple syrup and maple sugar are products from the juice of sugar maple (*Acer saccharum*); the US, Canada and Japan are the main producers. Maple syrup contains 98 % saccharides in its dry matter, sucrose contributing about 80–99 %.

2.1.6 Polyols (sugar alcohols)

As the name implies, polyols are the reduction products of mono-, di- and polysaccharides in which the carbonyl group has been reduced either to a primary or a secondary hydroxyl group.

As the hydrogenation of an aldose or a ketose produces sometimes an identical product, it follows that there are fewer saccharide alcohols than the parent sugars. Thus both D-glucose and D-fructose yield upon reduction D-glucitol (sorbitol).

Reduction can also increase the number of products from ketoses because an additional asymmetric centre has been created. Thus the reduction of D-fructose produces not only D-sorbitol but D-mannitol as well.

Ever since it was discovered that polyols can be used as a substitute for sucrose in the high-volume low-calorie sweeteners, they have received considerable attention. The extent to which they will displace sucrose eventually will depend a great deal on how successful they will prove to be in dietetic food and in non-cariogenic low-energy products.

2.1.6.1 Sorbitol, mannitol and maltitol

Sorbitol, an abundant naturally occurring polyalcohol (a mammalian metabolite) (Table 2.9) [116], was isolated for the first time more than

Table 2.9 — Occurrence of sorbitol in nature [116]

Fruit	Concentration (g/dm^3)
Apples	2.6–9.2
Pears	11.0–26.4
Cherries	14.7–21.3
Sour cherries	13.1–29.8
Plums	1.8–13.5
Peaches	0.7–1.2
Elderberries	0.17–0.28

a century ago. Mannitol is found in the sweet exudate from the ash tree (*Fraxinus ornus*). The exudate is called *manna* — hence mannitol. Mannitol is equally widely distributed in the plant kingdom, seaweeds and mushrooms are particularly rich. Maltitol is produced by the hydrogenation of maltose using Ni catalyst (Scheme 2.7), whilst the production of sorbitol starts with either sucrose, glucose or starch and that of mannitol with either fructose or mannose. Starch has recently emerged as the preferred starting material for the production of various polyalcohols [117] (Scheme 2.8).

The reduction of monosaccharides can be accomplished by various reagents such as sodium amalgam [118–120], metallic calcium [121] or catalytically using various rare metals [122]. Very good results have been achieved using sodium borohydride [123, 124]. Aldonic acid lactones can be also reduced to the corresponding alcohols [125, 126].

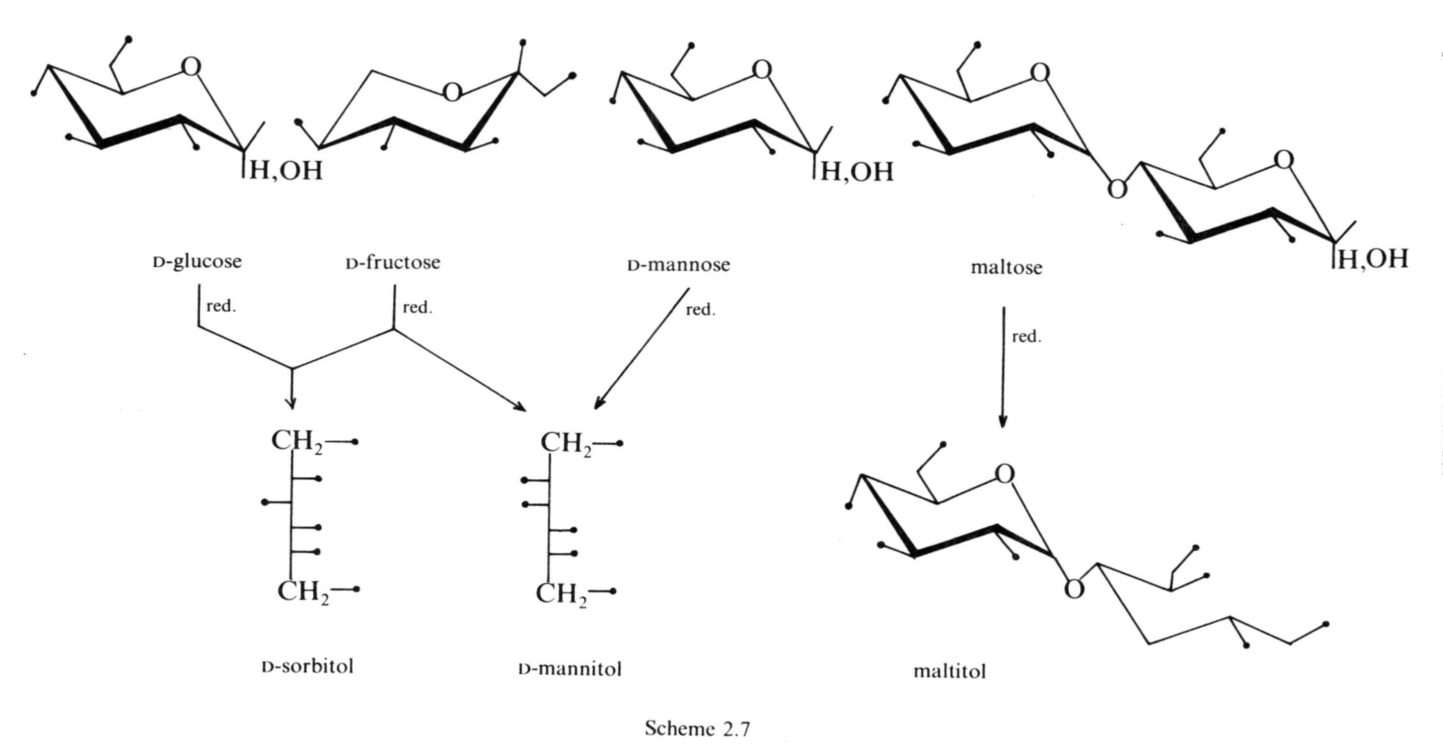
H,OH
H,OH
H,OH
D-glucose
D-fructose
D-mannose
maltose
red.
red.
red.
red.
CH_2—
CH_2—
CH_2—
CH_2—
D-sorbitol
D-mannitol
maltitol

Scheme 2.7

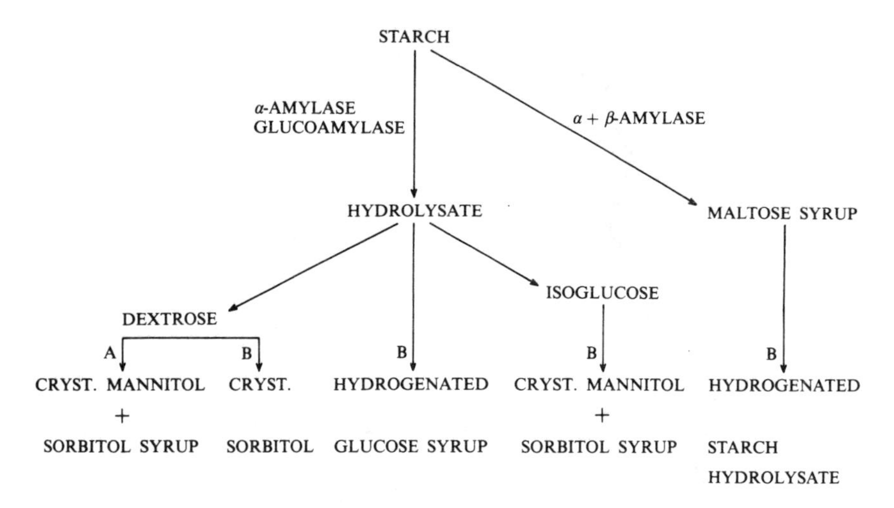

Reaction conditions :
A — $(NH_4)_2MoO_4 + (H_2 + RaNi)$,
B — $H_2 + RaNi$.

Scheme 2.8

D-Sorbitol and D-mannitol are prepared by hydrogenation and isomerization of glucose [127]. High-pressure hydrogenation of sucrose on ruthenium catalyst affords in the presence of phosphoric acid at 80 °C a mixture of 85–87 % of sorbitol and 12–15 % mannitol. Modification of the reaction conditions by the addition of optically active tartaric acid [128] and glutamic acid resulted in a 66 % yield of sorbitol.

Invert sugar is used preferentially as the raw material for the production of mannitol. Its hydrogenation yields a mixture of sorbitol and mannitol, which is separated most conveniently by fractional crystallization. Because of its limited solubility in water, most of mannitol separates readily and the mother liquor contains mainly sorbitol. The liquor is concentrated (70 % solids) and is used as such in various applications. If mannitol is the desired product, the hydrogenation is carried out under basic conditions [116].

Before starch can be hydrogenated into useful products, it must be hydrolysed to a syrup that contains glucose, maltose and higher saccharides. The next step is the nickel-catalysed high-pressure hydrogenation at elevated temperature which produces hydrogenated starch hydrolysate (HSH). The starch hydrolysis is carefully controlled in order to produce a syrup of clearly defined glucose and maltose content which in turn controls the sorbitol and mannitol relative concentrations in the hydrogenation liquor. This variation is responsible for a range of sweetness of the HSH, between 25 and 50 % relative to sucrose.

Malbit® is brand name of the product that contains over 90 % maltitol and less than 5 % maltotriitol. A contemporary version of HSH is Lycassin® with a similar composition (Table 2.10).

Table 2.10 — Composition maltitol-containing products

Component	Malbit® crystalline	Malbit® liquid	Lycasin® 80/85
Glucitol	3.8	1.3	6.7
Maltitol	88.4	79.9	53.8
Maltotriitol	5.6	12.1	16.3

The physico-chemical properties of polyols are summarized in Table 2.3. In comparison with the parent saccharides, the polyols have higher (endothermic) heat of solution and therefore produce a cooling sensation in the mouth. They do not undergo Maillard reaction because of the absence of free carbonyl group and they also resist microbial degradation better than sugars. Sorbitol, maltitol and HSH are very hygroscopic.

Metabolism of sorbitol has been studied thoroughly and the topic is covered in several outstanding review articles [117, 129–131]. The main mammalian metabolic pathways involve the oxidation by a dehydrogenase to fructose and further metabolic conversions *via* the fructoso-1-phosphate [129]. This route is the same for both D-mannitol and D-sorbitol [132].

Experiments with ^{13}C-labelled mannitol revealed that, although over 65 % of mannitol was absorbed, 17.5 % was recovered in the urine, suggesting that only about 50 % of ingested mannitol was utilized by the body. This metabolism translates to a very low calorific value for mannitol (Table 2.3).

The calorific utilization of maltitol in humans is also about 50 %. A portion is hydrolysed in the stomach into its component units — sorbitol and glucose.

Sorbitol has been used as a sweetening agent for diabetics since the late 1920s because it causes only a marginal elevation of blood sugar level [133, 134].

Some health professionals have discouraged the use of sorbitol by diabetics on the ground that it can be converted to glucose, and therefore it is bound to need insulin for its metabolism. These fears have been dispelled to some extent by the fact the sorbitol ingestion results in lower blood sugar levels. Thus sorbitol is expected to reduce the severity of diabetes complications brought on by the cellular accumulation of sorbitol [135] — a fact that is directly related to insulin deficiency induced hyperglycemia.

Frostell *et al.* [136] found that sorbitol is also suitable for the diet of diabetic children.

Mannitol is not very desirable in a diabetic diet because of its very low laxative threshold. Maltitol and MSH, which are converted enzymatically in the intestine into glucose and sorbitol, are also less suitable for the diabetic diet.

Sorbitol has been used in hard and soft candies, cough drops, jams, chewing gums, jellies, baked goods and baking mixes, in frozen dairy desserts and mixes, as well as in other foods. Foods that are expected to be consumed in quantities that would result in a daily ingestion of 50 g of sorbitol must carry a warning of their laxative effect.

2.1.6.2 Xylitol

Xylitol is a five-carbon sugar alcohol with a sweetness comparable with that of sucrose [137]. It is found in small amounts in a variety of fruits and vegetables [138] and is formed as a normal glucose metabolite in the human body [139]. Xylitol has been used as a sweetening agent in human food since 1960.

Xylitol was first described and synthesized by Fischer [140]. Commercial production of xylitol is based on the chemical conversion of xylan [141]. The most common sources of the latter are birchwood, straw, corn cobs, almond shells and waste paper from the paper and pulp industry. The production technology involves the hydrolysis of the disintegrated xylan containing raw material into xylose. Xylose is then isolated and converted into xylitol by nickel-catalysed hydrogenation. Alternatively, the impure xylose solution is hydrogenated and the resulting xylitol syrup is purified (Scheme 2.9). Other methods of synthesis [142, 143] that also include fermentation of glucose [144] have not been yet commercialized.

Xylitol is a white crystalline compound (Table 2.3) readily soluble in water. In common with other polyalcohols, it does not react in the Maillard reaction.

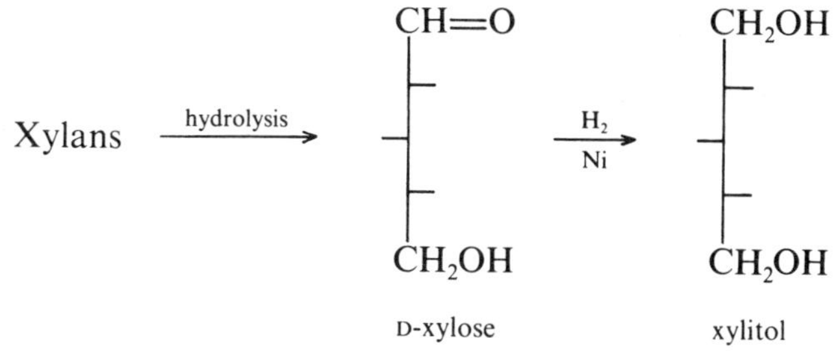

$$\text{Xylans} \xrightarrow{\text{hydrolysis}} \begin{array}{c} CH{=}O \\ | \\ \vdash \\ \dashv \\ \vdash \\ | \\ CH_2OH \end{array} \xrightarrow[\text{Ni}]{H_2} \begin{array}{c} CH_2OH \\ | \\ \vdash \\ \dashv \\ \vdash \\ | \\ CH_2OH \end{array}$$

D-xylose xylitol

Scheme 2.9

Xylitol metabolism has been formulated in terms of either direct metabolism of the absorbed xylitol by the mammalian organism [145–147] mainly after the liver, or as an indirect metabolism involving enzymatic degradation of unabsorbed xylitol in the intestine [148–150]. Xylitol is stable to degradation by oral microorganisms [151–154], i.e. the decrease of pH does not take place [151–159]. Similarly, tests of xylitol's acute toxicity, embryo toxicity, teratogenicity, mutagenicity and clustrogenicity were all negative [160].

Xylitol is being used as a non-cariogenic sweetener for confectionery (chewing gum, chocolate, hard candies) as well as in galenic preparations and in cosmetics (tooth paste, mouth spray, mouthwash) [161–163]. It is often used as a substitute for glucose in infusion solutions. Xylitol is not very suitable for soft drinks because of its mild laxative effect, and because it acts as an inhibitor of yeast activity, it is unsuitable as a sweetener in yeast-containing product [164].

2.1.6.3 Lactitol

Lactitol has been known for a long time but its sweetness has attracted attention only quite recently. It is a typical representative of the family of polyalcohols and has many of their typical properties.

Studies of lactitol were started around 1975 by CCA 'Biochem b.v.' a Dutch chemical company that produces sorbitol and mannitol, in cooperation with a fast growing producer of lactose, Whey Product Borculo.

The structure of lactitol (4-*O*-β-D-galactopyranosyl)-β-D-glucopyranose (Scheme 2.10), can be derived from that of lactose.

Lactitol is found in the milk of all mammals. Its sweetness is about 1/6 that of sucrose. The most common source of lactitol is whey from

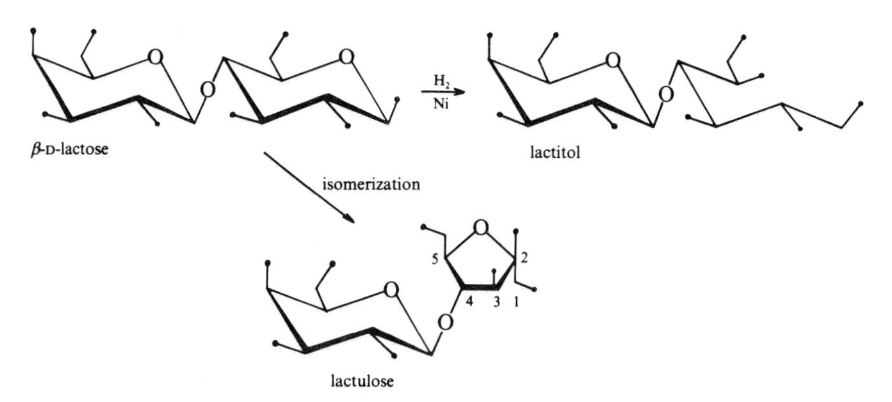

Scheme 2.10

which it is isolated directly by crystallization, sometimes the proteins can be removed by ultracentrifugation prior to crystallization. Table 2.11 gives a summary of the use of lactitol in food.

Lactose, the chemical precursor of lactitol, is poorly soluble in water (20 % solution can be made at RT); at higher concentrations it precipitates from the solution as hard crystals of the α-lactose hydrate which produce a sandy feeling in the mouth.

Table 2.11 — Sources of lactose in food (data in %)

Component	Lactose	Proteins	Fat	Ash	Moisture
Full fat milk	4.8	3.5	3.5	0.7	87.4
Skimmed milk	5.1	3.6	0.1	0.7	90.5
Sweetened condensed milk	16.3	10.0	0.3	2.3	28.4
Concentrated whey	38.5	7.0	2.4	4.0	48.1
Dried skimmed milk	52.0	35.9	0.8	8.0	3.0
Sweet dried whey	73.5	12.9	0.9	8.0	4.5
Dried sour whey	66.5	13.0	—	10.2	3.2
Whey solid matter	56.6	39.0	3.3	2.4	9.77

The low sweet potency of lactose can be improved by enzymatic treatment (lactase and glucoisomerase) of lactose syrup — this converts it to a syrup containing 10 % solids which contain 20 % fructose, 25 % glucose, 45 % galactose and 1 % lactose, in addition to a high content of salt.

Lactose has been used as a bulk carrier of intensive sweeteners in the food industry because it improves some of the quality parameters (browning of food in microwave ovens) and quality of milk drinks. However, a substantial part of the human population lacks the enzymes for digesting lactose.

The isomerization of lactose leads to lactulose, 4-*O*-(β-D-galactopyranosyl)-D-fructose, which is 1.5 times sweeter (Scheme 2.10); the reaction is catalysed by calcium hydroxide [165], alkali metal hydroxides [166, 167], magnesium oxide [168], anionic ion exchange resins [169] and amines [170]. Lactulose is utilized in the food industry, the pharmaceutical industry and in the production of animal feed.

Lactitol differs from lactose by the fact that the D-glucopyranose residue is reduced into D-sorbitol. This is achieved by a nickel-catalysed hydrogenation (Scheme 2.10), a reaction that was used as early as 1912 to produce lactitol syrup [171], crystals of lacticol dihydrate were obtained later [172, 173]. Industrial manufacture of lactitol involves nickel-catalysed hydrogenation of lactose solution, isolation and purification by activated carbon, ion exchange, evaporation and crystallization [174].

Lactitol hydrate is a colourless, sweet, non-hygroscopic, stable and free-flowing material with good handling and storing properties. It is softer than lactose and does not elicit a sandy feeling in the mouth. The viscosity of aqueous solutions is comparable to sucrose solutions of the same concentration. Furthermore, the crystallization behaviour of lactitol exhibits the same pattern as sucrose, an important factor to consider in production of confectionery items.

Lactitol is not absorbed as a whole molecule and small intestine enzymes, such as β-galactosidase, do not cleave it. Therefore, it is very well tolerated by diabetics [175]. Furthermore, it does not influence the blood glucose level and thus it is suitable for both the insulin and non-insulin-dependent diabetics.

Lactitol has a reduced calorific value and is metabolized in the large intestine in a similar way as dietary fibre.

The physiological and physical properties of lactitol qualify it as a suitable replacement for sucrose, both as a sweet and bulking agent (texturizing) in various products [175] such as chocolate, hard and soft candies, chewing gum, jam and marmelade, ice cream and frozen dairy dessert.

2.1.6.4 Palatinit

Palatinit (isomaltitol) is a sweetener whose characteristics are comparable to those of sucrose [176]. The obtained product is optically active, non-reducing and sweet. However, unlike sucrose, it is extremely stable towards enzymatic hydrolysis and cannot be fermented by the majority of yeasts and microorganisms prevalent in nature.

In fact, palatinit is a mixture of two isomers, 6-*O*-(D-glucopyranosyl)-D-sorbitol (GPS) and 1-*O*-(D-glucopyranosyl)-D-mannitol (GPM), both components being crystalline (Table 2.12).

Table 2.12 — Characteristic properties of palatinit [176]

	Palatinit	GPS	GPM
Molecular formula		$C_{12}H_{24}O_{11}$	$C_{12}H_{24}O_{11} . 2H_2O$
Molecular weight		344.3	380.3
m.p. (°C)		165–168	162
Temp. of agglomeration (°C)	145–150	—	—

The preparation of palatinit entails two steps [177] (Scheme 2.11). In the first step sucrose is enzymatically transglucosylated to isomaltulose — palatinose (6-*O*-α-D-glucopyranosyl-D-fructofuranose). The disaccharide linkage is shifted in the course of transglucosidation and thus results in the formation of a reducing sugar from a non-reducing one. The disaccharide isomaltulose crystallizes with one molecule of water of

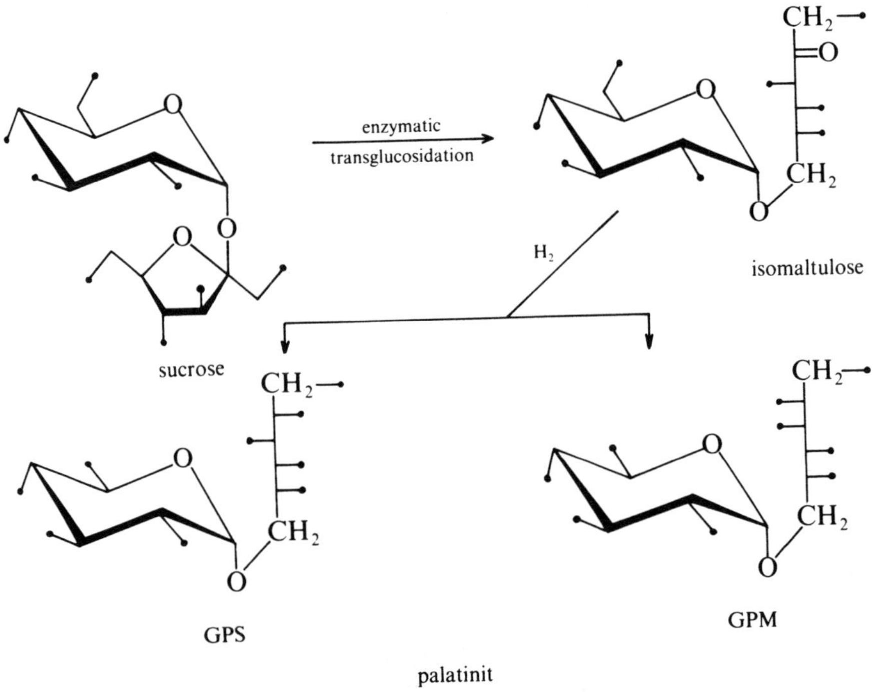

Scheme 2.11

crystallization, it is not toxic and occurs in nature (honey). The second step involves Raney nickel hydrogenation of isomaltulose into an equimolar mixture of GPM and GPS in a neutral aqueous solution (pH 6–8). Whilst GPM crystallizes with two moles of water, GPS yields anhydrous crystals. Thus crystalline palatinit contains approximately 5 % of crystallization water.

The two glucosyl hexitols in palatinit exist in different conformations. Whilst the GPS molecule is shorter and more voluminous, the GPM molecule has an elongated shape [178, 179]. The difference in molecular shape is also reflected in different metabolic pathways along which the molecules encounter disaccharidases [180], excretion by liver [180] and demonstrate their enzyme inhibiting properties [181]. The observed differences between them are rather subtle and afford no advantage for either.

Both GPS and GMP are very stable compounds; they do not react with other food ingredients, neither do they undergo the Maillard reaction. The increased stability is due to the more stable bond between the α-D-glucopyranosyl residue and the hexitol. Although the α-D-glucosidase should in principle be capable of cleaving both GPS and GPM, practical experiments [178] have shown that the rate of such enzymatic cleavage was three orders of magnitude smaller than that for maltose.

The disaccharide bond in GPS and GPM is also quite stable to acid and enzymatic hydrolysis. Palatinit can participate in microbial processes only after GPS and GPM have been broken into their components (glucose, mannitol and sorbitol). Thus palatinit cannot be used as a substrate for most yeasts, molds and bacteria that are found in food. Palatinit sweetened drinks and food have exceptional microbial stability. Other significant features of palatinit are its low cariogenicity [182–184] and lower calorific value [180, 185, 186]. Clinical tests on the effects of palatinit have revealed that it causes only insignificant changes in serum glucose and serum insulin [180, 187–191].

Palatinit is barely absorbed in the body. Its slow hydrolysis is accompanied by partial absorption of its metabolites — glucose, mannitol and sorbitol — in the small intestine [178]. Numerous animal experiments and clinical studies indicate a net reduction in calorific utilization of palatinit [176, 180]. The tolerance to palatinit has been closely monitored and experiments were designed to assess this aspect. The main conclusion of these studies is that the relative tolerance to palatinit and other sugars depends strongly on the form of the product in which it is ingested [176]. In this respect the intestinal tract is less sensitive to solid food than to liquid food. It has also been established that test animals adapt significantly faster to high doses of palatinit than to similar doses of other sugar alcohols. One of the reasons for this appears to be the higher molecular weight, and thus correspondingly lower osmotic pressure, of the disaccharide palatinit in comparison with monosaccharides. Prolonged ingestion of high doses of palatinit by experimental animals does cause neither more frequent diarrhoea and flatulence, nor metabolic disturbances. Finally, palatinit was found to be tolerated by non-adapted humans at least as well as comparable doses of other popular sugars.

Palatinit has a pure sweet taste that is similar to sucrose (0.45–0.65 sucrose) and no aftertaste. When mixed with aromatic foods, it potentiates their flavour. Similar synergistic effects occur when palatinit is mixed with other sugar alcohols. When used with intensely sweet artificial sweeteners such as saccharin or cyclamate, palatinit masks the unpleasant metallic aftertaste of these sweeteners and makes products in which this mixture is used difficult to distinguish from those sweetened with sucrose.

The product application studies [176] have been conducted in the following food categories:

candies	jellies, jams
chocolate	fruit-based preparations
praline chocolates	dough and pastry fillings
marzipan	cakes, other fine biscuits
chewing gum	pastry frosting and icing
ice cream	whipped cream
coffee/tea sweeteners	pudding and desserts
	refreshment beverages

2.1.6.5 Prospects for the production and consumption of polyols

There have been problems connected with the use of sugar alcohols as sweeteners in spite of the large volume of knowledge that is documented by many original papers and review articles.

One such notorious problem concerns with the effect of long-term use on the human organism. The rejection by some organisms after continuous use, due to intestinal discomfort, have caused concern about acceptance of sweeteners that contain polyols. However, there exists a great degree of variability among humans; even otherwise healthy individuals may be unable to adapt to certain types of sweeteners. In order to overcome this subtly varying tolerance, researchers have been encouraged to design a correspondingly broad spectrum of sweeteners with finely tuned properties in order to cater for such individual needs.

Future developments in polyols and other bulk sweeteners are expected to concentrate less on modified monomeric compounds than on modified glycosidic bonds (second generation of sugar substitutes). The once cherished endeavour to find a universal sweetener, suitable for any human, has been abandoned.

2.1.7 L-Saccharides

The possibility, and indeed the intent, to use L-sugars as sweeteners was announced in 1981 [192]. L-Sugars were rather neglected until then in the scientific literature [191, 193], even though they were considered in the public media.

L-Sugars are much rarer in nature than their D-form counterparts. They have been variously reported as minor unquantified constituents in a variety of natural products, such as:

L-fructose and L-rhamnose in plantain seeds
L-galactose in flax seed gum, red algae, snails eggs
L-arabinose in polysaccharide araban in sugar beets, mesquite, pectins
L-sorbose in berries of mountain ash and other plants
L-fucose in seaweed

Until recently, L-sugars were destined for research only. They were made by the resolution of D,L-mixtures by feeding them to bacteria which consumed only the D-form [194–203]. L-Sorbose, an intermediate in the synthesis of ascorbic acid, was nominated as a non-cariogenic bulk sweetener [191, 204], however, its chemical instability limits its wider use [205]. Its metabolism was thoroughly studied [206–209] and its metabolic [210] and other effects described [205]. As a consequence, this L-ketohexose was abandoned as a sucrose substitute in food. However, in other applications good use can be made of the fact that *S. mutans* NCTC 10449 ferments L-glucose anaerobically, and thus opens new horizons for its use as a sugar substitute [191].

Biospherics Inc. in the USA announced in 1987 that it prepared sufficient amounts of one L-sugar (Lev-O-Cal) to start tests, as required by FDA, in order to gain approval for food applications. The tests were expected to end by 1990 and the FDA approval was due in 1991 [211]. L-Fructose and L-glucose are contemplated for industrial production. Current estimates indicate that even at present-day costs a very profitable production of L-sugars is possible, for less than $ 1.00 per pound.

Production of L-sugars has been patented also in Czechoslovakia [212].

2.2 OTHER NATURAL SWEET SUBSTANCES

Although sucrose is the sole most important, popular and used sweet substance in the human diet, it does not make life sweet for everybody. Diabetics are the largest and the best known group of individuals whose need for an alternative sweetener had stimulated the search for other natural sweet compounds which could substitute sucrose in low-calorie and diabetic products.

A great number of sweet natural organic compounds have been found, some of which are already available commercially as sucrose substitutes. Naturally sweet organic compounds comprise many families (proteins, terpenoids, steroids, dihydroisocoumarins, flavanoids) of structurally unrelated compounds and with equally variable sweet potency. Glycyrrhizin, stevioside, thaumatin and phyllodulcin are some of the compounds that are being currently used in human diet as sweetening or flavouring agents. Many substances from this group of sweet compounds have been used in the sweetness–structure related studies.

2.2.1 Proteins

2.2.1.1 Thaumatins

The fruit of *Thaumatococcus daniellii* BENTH (family Maranthaceae) contains a very sweet tasting substance with a liquorice aftertaste. The plant, first described by Danielli [213], is grown in Western Africa from Sierra Leone, where it is called 'katemfe', to the Republic of Zaire. It also belongs to the indigenous flora in Uganda and Sudan; it is grown in hot-houses in England.

The fruit is used by the African natives to mask the acidity of palm wine and fruit drinks. The *Thaumatococcus* plant has broad, oval leaves with paper-like texture that are carried on 3 m long stems. A single fruit that weighs between 9 and 40 grams (average about 16 g) consists of 2–3 triangular fleshy pericarps which are red when ripe. The fruit contains large black seeds which are surrounded by a transparent viscous mucilage with a soft, jelly-like aril at the base. Inglett and May [214] established that the water-soluble sweet principle is contained in the aril, a membranous sac. Van der Wel and Loeve [215] isolated and characterized it, and named it thaumatin I and II. Starting with one kilogram of fruit flesh, they isolated, after ultrafiltration followed by gel filtration and ion exchange chromatography, 0.9 g of sweet protein with 1600 times the sweetness of sucrose (related to 7 % sucrose solution). This means that 0.9 g of thaumatins produces a sweetness equivalent to 1.5 kg of sugar.

Not surprisingly, these findings stimulated an extensive research into an improvement in isolation and structure elucidation of thaumatins. The yield was improved when the isolation was done in the presence of salts at low concentration [216, 225]. The aluminium salts proved to be specially suitable for this purpose; furthermore the product had better colour, stability, filterability, purity, and thus a higher quality of sweetness. The improved isolation procedure yielded 6 g of thaumatins from a kilogram of fruit.

The proteinaceous character of the two sweet tasting compounds was proven by the characteristic UV absorption spectrum. Thaumatin I and II are very similar, as can be seen from their amino acid composition (Table 2.13) which is practically identical. Both molecules have the same *N*-terminal amino acid alanine and have molecular weights of 21,000 + 600 and 20,400 + 600, respectively, isoelectric points $T_I = T_{II} = 11.7–12$ [215, 216].

Table 2.13 — Number of amino acid residues in thaumatins [215, 225]

		Thaumatin I		Thaumatin II	
Aspartic acid	Asp	21	12	19	13
Glutamic acid	Glu	10	6	10	6
Serine	Ser	12	14	10	13
Threonine	Thr	19	20	17	20
Proline	Pro	12	20	12	20
Glycine	Gly	23	24	22	24
Alanine	Ala	15	16	14	16
Valine	Val	9	10	8	10
Leucine	Leu	9	10	9	10
Isoleucine	Ile	7	8	7	8
Methionine	Met	1	8	1	8
Phenylalanine	Phe	10	11	10	11
Tyrosine	Tyr	7	8	8	8
Cystine	Cys	14	16	13	16
Lysine	Lys	10	11	11	11
Arginine	Arg	11	12	12	13
Tryptophan	Trp	3	12	3	3
Asparagine	Asn		10		8
Glutamine	Gln		4		5

Careful fractionation yielded three other sweet proteins designated T_a, T_b, and T_c which were eluted from the column before T_I and T_{II} [216, 217]. The three new proteins have the amino acid content and the molecular weights similar to those of thaumatins but are less sweet.

The amino acid sequence of T_I was first described in 1979 [218]; T_I

consists of a single polypeptide chain of 207 amino acids and eight intramolecular disulphide bonds. The calculated molecular weight, based on the known amino acid sequence, is 22,209.

The amino acid sequence of T_{II} was determined later by methods of genetic manipulation [225] — it is identical to that of T_I except for five different units at positions 46, 63, 67, 76 and 113. The calculated molecular weight is 22,293.

The disulphide linkages of the 16 cystine residues in thaumatins have been investigated by van der Wel *et al.* [219] who were able to assign the disulphide bonds to cysteines in positions 9–204, 56–66, 71–77, 121–193, 126–177, 134–149, 145–158 and 159–164.

Proteolytic digestion of thaumatin I and partially S-[^{14}C]carboxymethylated thaumatin I was performed with trypsin, chymotrypsin and pepsin. It was found by sequential analysis of the amino acids that the labile disulphide bond that is responsible for the enzymatic properties of thaumatin connects Cys-145 and Cys-158.

Chemical (oxidative) decomposition or reductive cleavage leads to the total loss of sweetness [220]. The dominant factor for the sweet taste sensation is the tertiary structure as indicated by previous psychophysical, chemical and spectroscopic evidence. In spite of the known primary structure and the position of the network of the disulphide linkages which are responsible for holding the protein chain in the correct conformation, the exact role of the three-dimensional folding of the thaumatin molecule in eliciting the sweet sensation remains as yet largely unknown [224].

Crystal structure studies have shown that thaumatin has three structural domains. The largest one is a flattened β-barrel formed by 11 antiparallel β-strands, each being, on average, six residues long [222]. CD (circular dichroism) measurements and theoretical calculations helped to identify two regions in the molecule of thaumatin (Fig. 2.11) each possessing a small α-helix, in addition to a certain number of β-structures and 4 disulphide bridges [221].

Van der Wel described in detail selective chemical modification of amino acids (Lys, Arg, Try, Met, Asp, Cys) in the thaumatin side-chain as well as the concomitant changes in sweetness [224, 225]. He also showed [220] that as more lysine residues were acetylated, the sweetness of T_I decreased, so much so that when 4 out of the 10 lysine units were acetylated, the sweetness disappeared completely.

Temperature affects the sweetness of thaumatins in a rather complex manner. The effect depends on the concentration, presence of oxygen, salts or polyelectrolytes and on pH.

Korver *et al.* [223] showed that irreversible thermal denaturation

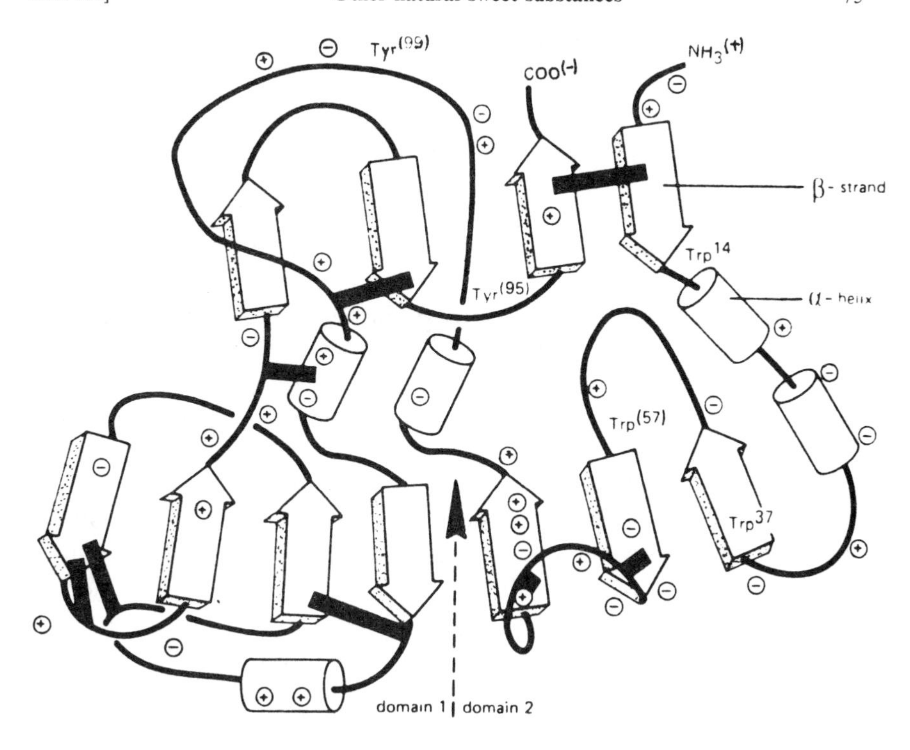

Fig. 2.11 — Packing diagram of the natural protein — thaumatin I.

(heat induced conformational change) occurred at 75° C and pH 5, again at 55° C and pH 3.0 and also at 65° C and pH 7.2.

Aluminium cation affects the sweetness of thaumatin profoundly [216, 217]. Ionic adduct thaumatin-Al, called Talin by Tate & Lyle PLC (UK), is a very powerful sweetening agent, 3000 times sweeter than a 10 % sucrose solution. It is extremely soluble in water, soluble in ethanol, propanol-2, glycerol and propylene glycol; it is insoluble in acetone, ether and toluene.

Aqueous solutions of Talin are stable in the pH range of 2 to 10. Thermal stability is affected mainly by pH, presence of oxygen and other soluble matter, specially polysaccharides. Talin can be pasteurized or UHT sterilized (ultrashort heating to high temperature) in non-alcoholic beverages in the pH range of 2.8–3.5.

Microbial synthesis of thaumatin by the recombinant DNA techniques was described in 1985 by Edens and Van der Wel [221].

Possessing an outstanding combination of favourable properties, Talin has every chance to become the sweetener of choice in the production of chewing gums (50–150 mg of Talin per kg) and sweetening of coffee and tea. When used in soft drinks, 10–15 mg of Talin can substitute 20–30 % of sucrose. It is also suitable for use in gelatin-based confectionery and in tobacco flavours. Talin successfully masks the unpleasant taste of drugs in medicinal applications.

Talin is currently permitted to be used in Japan, UK [226] and Australia [227], and some regulatory status has been also introduced in other countries and organizations [225]. Thaumatins are also combined with saccharin and L-glucose, with xylitol and others [225]. Thaumatin admixed with glycyrrhizin and amino acids is marketed in Japan as San Sweet T-1 and T-100.

Talin, when used either with sucrose, or Acesulfam K or stevioside, showed some degree of synergism, but none was observed with cyclamates and aspartame.

Talin showed no acute toxicity in dog and rats up to doses of 20 g per kg of body weight. Subacute toxicity was studied over a 90-day period and resulted in the finding that up to 8 % of Talin can be administered to animals in food. Talin is neither teratogenic nor cariogenic [224].

Intense sweetness, non-cariogenicity, favourable gestatory properties predestine thaumatins for their role as novel food additives [224].

2.2.1.2 Monellin

The intensely sweet tasting berries of the West African plant *Dioscoreophyllum cumminsii* from the family Menispermaceae were described for the first time in 1895 [217], Irving described them in greater detail in 1961 [228]. The plant is indigenous to tropical western Africa, Sudan, equatorial Africa as well as to Zimbabwe, Mozambique and Kenya.

D. cumminsii is a climber with long twinning stems and heart-shaped leaves. The berries are red, about half an inch in diameter, and grow in grape-like clusters with about 50 to 100 berries in each bunch.

The sweet proteins are isolated in a similar manner to thaumatins, by homogenization with water of the whitish mucilaginous pulp obtained from the fruit, followed by centrifugation. Further purification procedure involves ion-exchange chromatography, desalting by ultrafiltration and then freeze-drying.

The first attempts to elucidate the structure were made by Inglett and May [229] who suggested that it was a high molecular weight (m. wt. ca 10,000) saccharide and assessed its sweetness to be 1600 times that of sucrose. Two independent research groups, from Monell Chemical Senses Centre (Morris and Cagan) [230] and Unilever Research (Van der Wel) [231] reported shortly thereafter that the material was in fact a protein, named monellin. Contemporary extraction techniques yield about 3 to 5 grams of pure protein from one kilogram of fruit. Monellin is 1500–2000 times sweeter when assessed against 7 % solution of sucrose.

The material isolated by Morris *et al.* [232] has a sweetness 3000 times that of sucrose on a weight basis, a molecular weight of 10,700 daltons and an isoelectric point (IEP) at pH 9.3. Van der Wel and Loeve [233] reported a molecular weight of 11,500 daltons and an IEP at pH 9.03. They also reported somewhat lower sweetness — 2000 to 2500 times that of sucrose.

Amino acid sequence analysis revealed that monellin was built up from two dissimilar, tightly and non-covalently bound polypeptide chains, designated as A and B, and composed of 44 and 50 amino acids, respectively [234, 235] (Tables 2.14, 2.15).

Table 2.14 — Composition of monellin

Amino acid		Chain A	Chain B	Natural monellin
Aspartic acid	Asp	4	6	10
Glutamic acid	Glu	4	8	12
Serine	Ser	2	—	2
Threonine	Thr	1	3	4
Proline	Pro	4	2	6
Glycine	Gly	3	5	8
Alanine	Ala	2	1	3
Valine	Val	2	2	4
Leucine	Leu	4	2	6
Isoleucine	Ile	2	6	8
Methionine	Met	—	1	1
Phenylalanine	Phe	2(3)	3	5
Tyrosine	Tyr	5	2	7
Lysine	Lys	4	5	9
Arginine	Arg	5	2	7
Tryptophan	Trp	—	1	1
Cystine	Cys	—	1	1
Total		44	50	94

Van der Wel [224] reported that the sweet material was a mixture, consisting mainly of monellin 4 and the minor monellins 1, 2, 3 and 5, the last having the amino acid composition 97/98, m. wt. 11,000, IEP at pH 9.0–9.4 and relative sweetness of 1200–3200 (sucrose = 1).

A molecular weight of monellin was calculated by Frank and Zuber [236] to be 11,069 when based on the known amino acid sequence. The

Table 2.15 — Amino acid sequence in monellin

Chain A:	
Phe-Arg-Glu-Ile-Lys-Gly-Tyr-Glu-Tyr-Gln-Leu-Tyr-Val-Tyr-Ala-Ser-Asp-Lys-Leu-Phe-Arg-Ala-Asp-Ile-Ser-Glu-Asp-Tyr-Lys-Thr-Arg-Gly-Arg-Lys-Leu-Leu-Arg-Phe-Asn-Gly-Pro-Val-Pro-Pro-Pro	
Asp : Asn = 3 : 1	Glu : Gln = 3 : 1
Chain B:	
Gly-Glu-Trp-Glu-Ile-Ile-Asp-Ile-Gly-Pro-Phe-Thr-Gln-Asn-Leu-Gly-Lys-Phe-Ala-Val-Asp-Glu-Glu-Asn-Lys-Ile-Gly-Gln-Tyr-Gly-Arg-Leu-Thr-Phe-Asn-Lys-Val-Ile-Arg-Pro-Cys-Met-Lys-Lys-Thr-Ile-Tyr-Glu-Asn-Glu	
Asp : Asn = 2 : 4	Glu : Gln = 6 : 2

authors also reported that the monellin molecule did not contain any histidine and only one cysteine.

Monellin was also obtained in a crystalline form [216] and its X-ray analysis showed that the monoclinic crystal contained 4 molecules in a unit cell [224]. Figure 2.12 shows the three-dimensional structure of monellin [222].

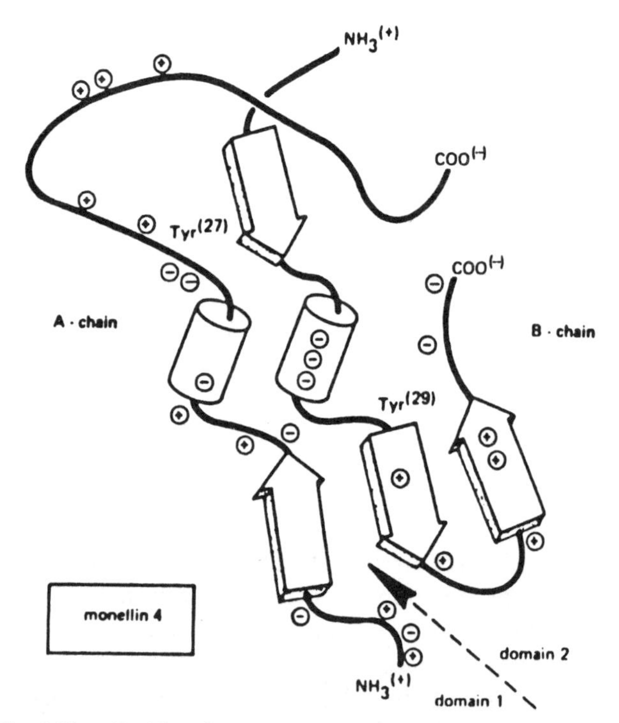

Fig. 2.12 — Packing diagram of the natural protein — monellin 4.

The conformation of the molecule of sweet peptides affects decisively the taste properties. The whole peptide molecule must have its original unperturbed conformation in order to elicit a sweet response.

Therefore, it should not be surprising that heating of aqueous solutions to 55–66 °C, or enzymatic hydrolysis of proteins, can eradicate the sweetness. Indeed, heating of a solution at pH 2 for half an hour at 70 °C results in total loss of sweetness. The sweetness also decreases in the pH range of 2–9 even at average room temperature. A complex of monellin and sodium alginate appears to be more stable towards heating at pH 2.1 than pure monellin.

As might be expected, the separation of monellin into its two constituent subunits A and B results in total loss of sweetness, neither A nor B alone are sweet.

Denaturation of monellin, with concomitant loss of sweetness, can be achieved also by treatment either with urea or guanidine hydrochloride. This process is accompanied by a distinct change in the emission fluorescence spectra. This observation led the authors [237] to suggest that emmission fluorescence spectra can serve as a sensitive probe of monellin's biological activity, which is in turn related to the conformation [238].

The secondary structure of monellin, as modelled by theoretical calculations, indicates the greater part of the molecule to be in a β-helix form, the remainder being in an α-helix.

Van der Wel described in detail the selective chemical modification of amino acids in the side-chain [224].

Monellin served as a model in studies related to structure–sweetness relationship, both in humans and several families of mammals [239]. The tests revealed that although the perception of sweetness in humans varied widely, it varied considerably more in mammals. It is interesting that some mammals did not respond to monellin even though their response to the sweetness of sucrose was normal. This suggests the need for receptors in specialized cells that are sensitive to proteins; these may be lacking in some mammals [216, 240].

Even though the toxicity studies are not yet completed, monellin is considered to be non-toxic. However, its thermal instability, sensitivity to pH and low abundancy in nature make it an unlikely candidate for a commercial sweetener.

Novel multiple (mixed) sweeteners were patented for the use in soft drinks [241]. Their preparation involved the dissolution of sweet peptides in water, an addition of weakly acidic polysaccharides (e.g. gum arabic) and followed by evaporation to dryness.

2.2.2 Terpenoids

The majority of intensely sweet naturally occurring compounds belong to the family of terpenoids. Many of the members are glycosides containing one or more saccharidic units which improve their water solubility. Natural sweet terpenoids comprise mono-, sesqui-, di- and triterpenoids.

2.2.2.1 Monoterpenoids

2.2.2.1.1 Perillartine — (*E*)-perillaldoxime

Perillartine (Fig. 2.13b) belongs to the group of chemically modified natural sweeteners; it is a sweet semisynthetic (*E*)-oxime of perillaldehyde (Fig. 2.13a), known since 1920 [242].

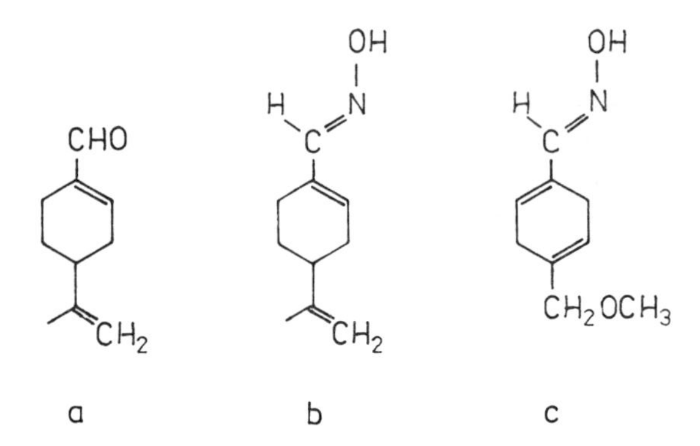

Fig. 2.13 — Structure of (a) perillaldehyde, (b) perillartine, (c) (*E*)-4-methoxymethyl-1,4-cyclohexadiene-1-carbaldoxime.

Perillaldehyde is itself a weakly sweet tasting monoterpene (12 times sweeter than sucrose), a component of the essential oil from the Chinese plant *Perilla frutescens* (L.) BRITTON (family Labiateae).

The structure of perillartine was determined by ^{1}H NMR studies [244]. In order to improve the solubility and remove the bitter taste, some 80 perillartine analogues were prepared [245]. Numerous structural analogues allowed the derivation of certain trends in the relationship between the structure and the desired properties as a sweetener; they are:

(a) most intense sweetness is achieved when the unsaturated oxime moiety assumes an E-α,β-configuration)

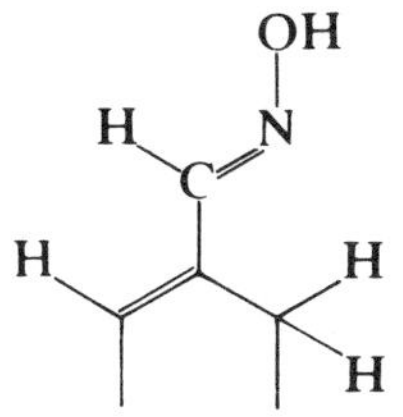

(b) polar groups in the side chain of perillartine improve water solubility but reduce the sweetness
(c) oxime ethers appear to be the derivatives which achieve the best balance between solubility and sweetness. An incorporation of an oxygen atom into the ring decreases sweetness
(d) alicyclic part of the perillartine molecule promotes sweetness whilst linear chains, bicyclics and aromatic rings reduce it, or even increase bitterness

The fusion of the above empirical rules led to the synthesis of (*E*)-4-(methoxymethyl)-1,4-cyclohexadiene-1-carbaldoxime V (Fig. 2.13c) which was found to be 450 times sweeter and better tasting than sucrose [246]. The sweetener was stable at pH 3 in the majority of foods and preparations.

The synthesis of the above compound was patented in 1975; its synthesis is shown in [247] Scheme 2.12.

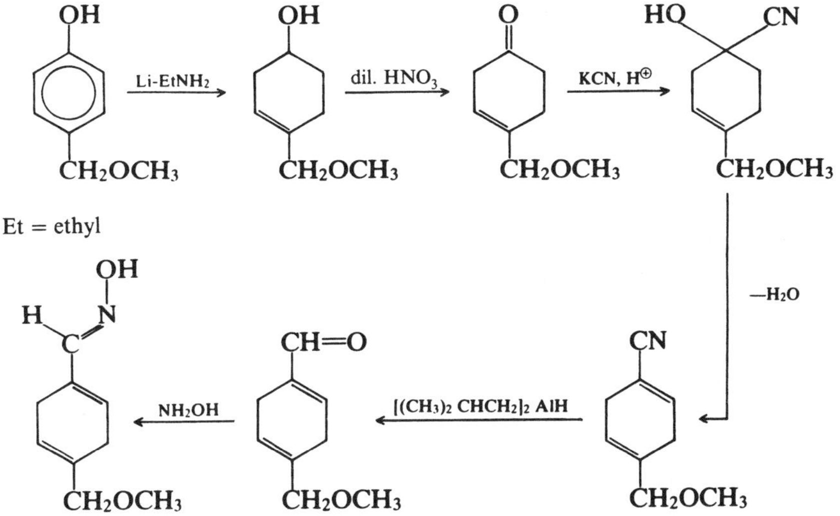

Et = ethyl

Scheme 2.12

Another seven-step synthesis that utilizes isoprene as the starting material was published in 1985 [248] and the product was used for the toxicological studies. Metabolism of the oxime V was examined with a ^{14}C-labelled compound; its synthesis started with 4-methoxybenzyl alcohol with specific activity of 3.51 mCi/mole and radiochemical purity of 98 %.

Several research groups investigated the structure–activity relationship of perillartine. Iwamura [249] utilized QSAR analysis of 49 perillartine analogues for this purpose. Italian authors [250] used perillartine analogues as models for producing two-dimensional electrostatic potential diagrams. They were used to assess the sweet and bitter response in humans.

Perillartine has been used to improve the taste of tobacco in Japan [243] and it was also used as a maple syrup and liquorice substitute [226]. Unfortunately, its usefulness as a sweetener is marred by its bitterness and poor water solubility even though it is 350 to 370 times sweeter than sucrose [226, 227].

Toxicity tests showed that oxime V (Fig. 2.13c) was not mutagenic and its LD_{50} of oral doses was 1 g/kg of body weight of male rats and 2 g/kg for male mice. Oxime V metabolizes fast and leaves the body easily, as has been demonstrated by the experiments that used ^{14}C-labelled substrates. Main metabolites arise from the oxidation and reduction of the cyclohexadiene ring and from the thiomethylation of the aromatic ring, as well as by the *O*-glucuronization of the aldoxime group [248].

2.2.2.2 *Sesquiterpenes*

2.2.2.2.1 Hernandulcin

Kinghorn and his co-workers isolated in 1985 an extremely sweet sesquiterpene from an Aztec herb *Lippia dulcis* Trev. (Verbenaceae). Aztecs knew the herb as Tzonpelic xihuitl (literally 'sweet herb'); it was also described in the book written between 1570–1576 by a Spanish physician Francisco Hernandez. The plant is found in central Mexico, especially around Mexico City. The sweet principle is isolated from the leaves and flowers by extraction, the extract is chromatographed on silica gel and then purified by thin-layer chromatography. The isolated sweet compound was named hernandulcin, in honour of Francesco Hernandez.

Hernandulcin (Fig. 2.14) is a colourless oil; its structure was inferred by Compadre and his co-workers [251] from IR, UV, MS, ^{1}H and ^{13}C NMR spectral data to be that of a sesquiterpene, 6-(1′,5′-dimethyl-

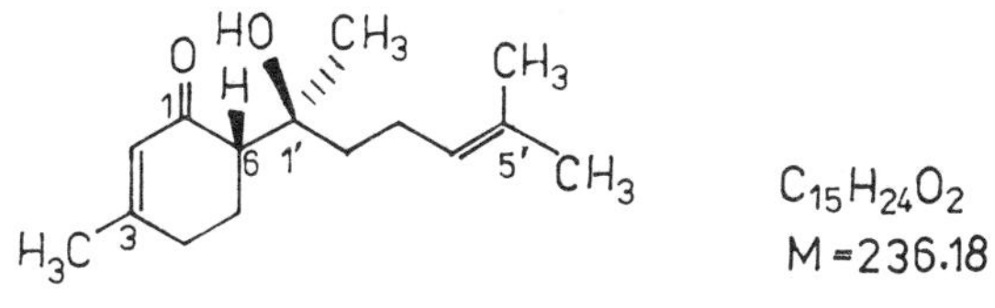

$C_{15}H_{24}O_2$
M = 236.18

Fig. 2.14 — Structure of hernandulcin.

1′-hydroxy-4′-hexenyl)-3-methyl-2-cyclohexanone (bisabolane). The structural assignment was confirmed by direct chemical synthesis which involved an aldol condensation between 3-methyl-2-cyclohexanone and 6-methyl-5-hepten-2-one in tetrahydrofuran (THF) at −15 °C (Scheme 2.13). (±)-Hernandulcin, similarly as (±)-epihernandulcin (6*S*,1′*R* and 6*R*,1′*S*), represents a 1 : 1 mixture of the (6*S*,1′*S*) and (6*R*,1′*R*) enantiomers [257].

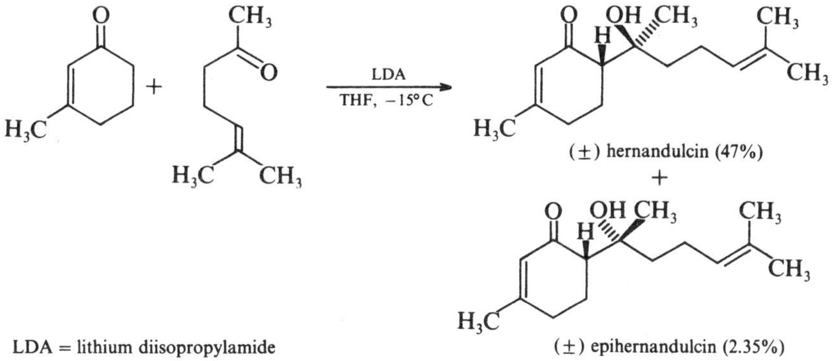

LDA
THF, −15°C

(±) hernandulcin (47%)
+
(±) epihernandulcin (2.35%)

LDA = lithium diisopropylamide

Scheme 2.13

Mori and Kato [253] prepared all four stereoisomers of hernandulcin from the enantiomers of limonene. The absolute configuration of the naturally occurring hernandulcin has been established to be 6*S*,1′*S*, the other isomers lacked sweetness.

The synthesis of natural hernandulcin started from *R*-(+)-limonene (Scheme 2.14). An efficient purification of the final product was achieved by overpressured layer chromatography [254].

(6*S*,1′*S*)-(+)-Hernandulcin has sweetness 1100–1200 times that of sucrose [253] but it lacks the pleasant, pure taste of the latter. It tastes fairly bitter, the other isomers are bitter and lack sweetness completely, and its solubility in ethanol is 0.11 g/100 ml. The modification of functional groups affects adversely the sweetness, the C_1 carbonyl and

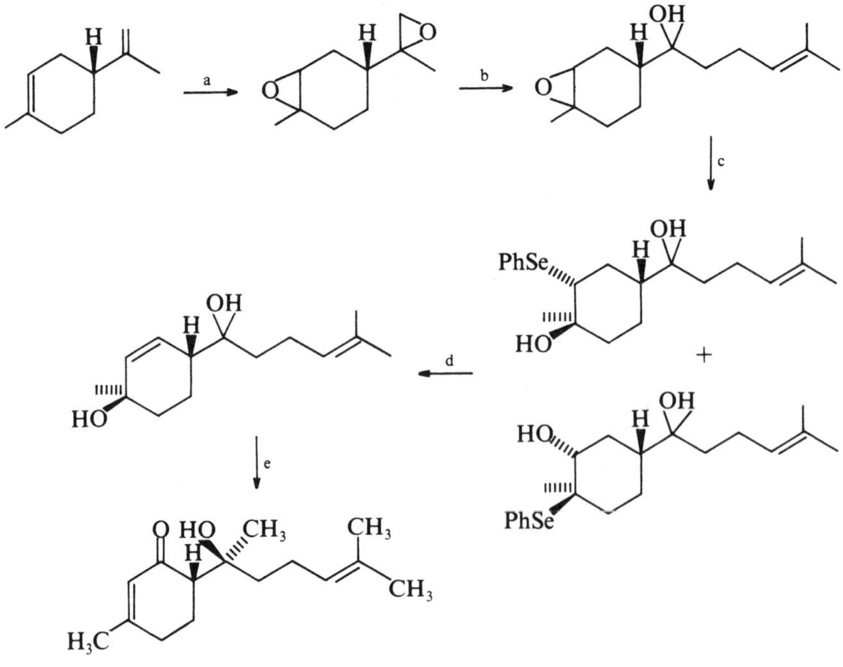

(6*S*, 1′*S*)-(+)-hernandulcin

a — chloroperbenzoic acid,
b — prenylmagnesiumchloride.
c — PhSeNa in EtOH, chromatographic separation,
d — H_2O_2 in THF,
e — $CrO_3 . C_5H_5N . HCl$ in CH_2Cl_2.

Scheme 2.14

the $C_{1'}$ hydroxyl groups are specially sensitive. Scheme 2.15 shows some chemical modifications of hernandulcin by acetylation and reduction.

Compounds a and b (Scheme 2.15) are not sweet. The reason is that according to [252], C_1 carbonyl and $C_{1'}$ hydroxyl groups constitute the Schallenberger AH–B system that is involved in the receptor–sweet substance interaction.

The toxicity of hernandulcin is low, acute toxicity was found to be 1 g/kg of body weight in mice. Tests with *Salmonella typhimurium* strain TM 677 proved also that it was not mutagenic [252].

In spite of its high sweet potency, hernandulcin is handicapped by its unpleasant aftertaste and natural bitterness for the use in many applications.

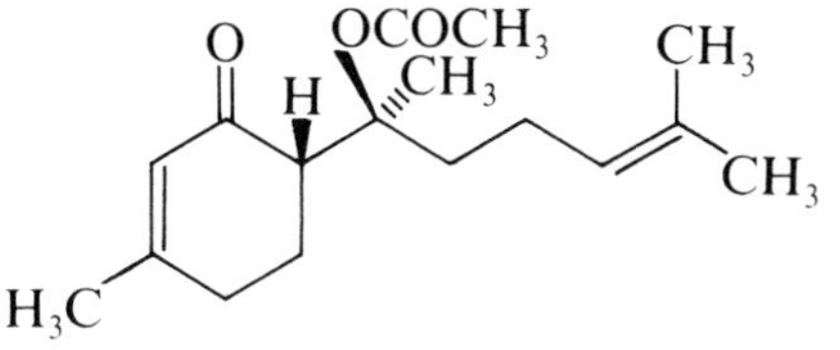

a 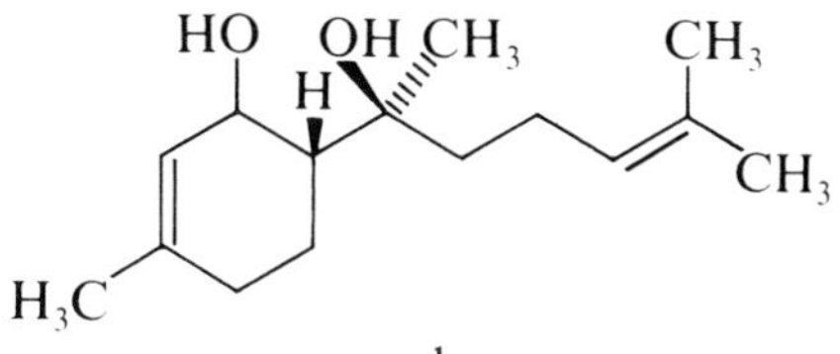 b

Scheme 2.15

2.2.2.3 *Diterpenoids*

2.2.2.3.1 Compounds from *Stevia rebaudiana* (BERTONI) (sweet *ent*-kaurene glycosides)

Stevia rebaudiana BERTONI (Compositae) is a sweet herb indigenous to the elevated terrain in the northeastern corner of Paraguay and to the neighbouring Brazil; it is called locally Yerba dulce. The plant is also grown in Japan, China and Korea.

Eight sweet *ent*-kaurene glycosides have been isolated from the dried leaves of *Stevia rebaudiana*, in addition to steviolbioside, rebaudiosides A–E and dulcoside A (Table 2.16). The total yield of the isolated sweet glycosides was 10 %.

Table 2.16 Structure of sweet *ent*-kauren glycosides

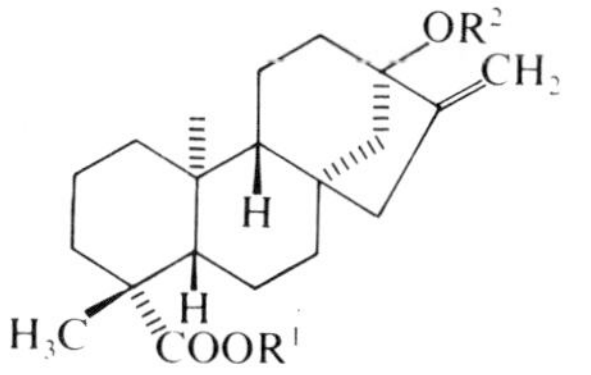

Compound	Trivial name	R^1	R^2
1	Stevioside	β-Glc	β-Glc^2-β-Glc
2	Steviolbioside	H	β-Glc^2-β-Glc
3	Rebaudioside A	β-Glc	β-Glc^2-β-Glc (β-Glc attached at 3)
4	Rebaudioside B	H	as A
5	Rebaudioside C	β-Glc	β-Glc^2-α-Rha (β-Glc attached at 3)
6	Rebaudioside D	β-Glc^2-β-Glc	as A
7	Rebaudioside E	β-Glc^2-β-Glc	β-Glc^2-β-Glc
8	Dulcoside A	β-Glc	β-Glc^2-α-Rha

Glc D-glucopyranosyl-, Rha L-rhamnopyranosyl-.

Plant extract and stevioside, its major constituent, are available commercially in Japan where they have been used for nearly a decade to sweeten soy sauce, sea foods, pickled vegetables, soft drinks and dessert items [226, 227]. Commercial products come in two purity grades 50 % and 90 % [225]. *S. rebaudiana* products have been cleared for sale in Brazil (e.g. a tabletop sweetener Stevita). Teas made from *S. rebaudiana* are prescribed by physicians in Paraguay for the treatment of the reported hypoglycemic activity. Frequently it is used in combination with other sweeteners, for instance with fructose in soft drinks, with various saccharides (sorbitol, xylitol) in chewing gums, or with glycyrrhizin. The sweetness of stevioside was found to be potentiated by aspartame, cyclamate and Acesulfam K but not by saccharin [225].

The isolation of stevioside from the dried leaves was described by Wood *et al.* [255]; the other seven sweet glycosides that were isolated [256] had their chemical structures confirmed by Kohda's group [257, 261].

The total yield of diterpene glycosides that were isolated from the dried leaves of *S. rebaudiana* of Chinese origin was determined by high-pressure liquid chromatography (HPLC) and found to be 6.6 % (stevioside), 3.7 % (rebaudioside A), 2.1 % (rebaudioside C) and 0.53 % (dulcoside A) [258]. An attempted separation by overpressured layer chromatography (OPLC) failed to yield any rebaudiosides D and E [254]. The isolation of stevioside can be carried out economically only from leaves; its content in flowers is far too low [259]. Patented isolation of stevioside and rebaudioside A from leaves claimed a total yield of 17.5 % (76.4 % stevioside and 1.4 % rebaudioside A) [260].

Stevioside (Fig. 2.15) is a white, crystalline hygroscopic powder with a melting point 196–198 °C. Its solubility in water is very good and it is also soluble in ethanol and propylene glycol. A tiny amount of stevioside, when put on the tongue, has a pleasant sweet taste, bigger amounts taste sweet at first, then a distinct bitter sweet taste comes through. The enzymatic hydrolysis of stevioside furnishes three moles of glucose and a mole of an insipid aglycone called steviol — *ent*-13-hydroxykaur-16-en-19-oic acid. Mossettig and his co-workers investi-

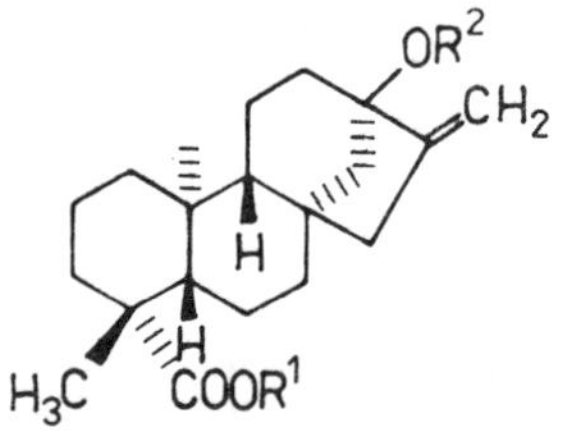

Fig. 2.15 — Stevioside R^1 = β-Glc, R = β-Glc-β-Glc (2 ← 1); steviol $R^1 = R^2$ = H. Glc — glucopyranosyl.

gated the structure of the aglycone and determined also its absolute configuration [262, 263]. They also established that all sugar residues were attached to the same side of the rigid aglycone.

Kotani [264] reported the isolation of steviol from stevioside (*S. rebaudiana* calus cultures) in 60 % yield. Chinese authors [265] were able to double the stevioside and rebaudioside A production in leaves and quadruple it in flowers.

Steviol (Fig. 2.15) could be prepared from stevioside chemically by the action of $NaIO_4$ and NaOH [266], or by enzymatic hydrolysis performed by the action of commercial pectinase [267].

Alkaline hydrolysis of stevioside and rebaudioside B furnishes steviolbioside and rebaudioside A [268].

Several synthetic steviolbioside derivatives have been prepared in order to assess the structure–sweetness relationship in diterpenoids (Table 2.17). It was found that increased branching of the group at C_{13} of the stevioside promoted pleasant sweet taste [273], but derivatives without the glycosidic unit were bitter [269]. Differences in the sweetness of steviosides (Table 2.17, derivative 3) depend on the method of assessment of the organoleptic properties, as well as on the concentration of the reference sucrose solutions [227].

Aqueous solutions of stevioside are stable within the pH range of

Table 2.17 The structure of partially synthetic steviolbiosides

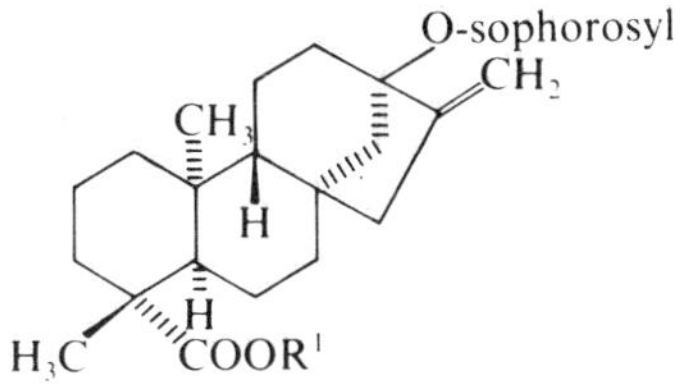

O-sophorosyl
CH_2
CH_3
H
H
H_3C
$COOR^1$

Compound	R^1	Rel. sweetness	Ref.
1	$(CH_2)_3SO_3Na$	160[b]	[269, 270]
2	$(CH_2)_2CH(SO_3Na)_2$	120[b]	[271]
3	β-Glc	255, 210[c]	[226, 272]
		190[b], 300[d]	[269, 270]
4	β-Xyl	160[a]	[272]
5	α-Ara	285[a]	[272]
6	α-Man	285[a]	[272]
7	α-Rha	200[a]	[272]

a – sucrose = 1, b – 10% solution of sucrose, c – 0.6% solution of sucrose, d – 0.4% solution of sucrose.

β-Glc – β-glucopyranosyl, β-Xyl – D-xylopyranosyl, α-Ara – L-arabinopyranosyl, α-Man – D-mannopyranosyl, α-Rha – L-rhamnopyranosyl.

3–9; they preserve their sweetness even after boiling for 1 hour. Stevioside solutions lose 50 % of their sweet potency in strongly basic conditions (pH > 10) [225]. Thus, for instance, a 1.8 % solution of lactic acid containing 0.1 % stevioside failed to show any signs of degradation and kept its sweetness even after storage at 80 °C for 5 hours. Stevioside is stable towards other components in food and does not give rise to browning. It can therefore be safely assumed that the processes that are involved in food preparation and food storing will not degrade stevioside to steviol [274].

Stevioside appears to be fairly innocuous, it is neither toxic [275] nor mutagenic (tested against *Salmonella typhimurium*) in the presence or absence of the metabolic activating system. Structurally related isosteviol and diterpenoids such as rebaudiosides and dulcoside were not active either, regardless of metabolic activation [275].

However, steviol, the aglycone of stevioside, was found to be highly mutagenic when evaluated in the presence of metabolic activators [276]. A concentration of 2.5 mg/ml elicited a similar response as an exposition to that of 80 μ*M* of benzo[*a*]pyrene or 150 μ*M* of 2-acetylaminofluorene [274]. The double bond was recognized as being responsible for the mutagenicity. A potential metabolite of steviol, steviol-16*α*,17-epoxide, as well as dihydrosteviols (prepared by catalytic hydrogenation on 5 % Pt-C), were found to be ineffective as direct-acting mutagens. Thus, although stevioside itself appears to be innocuous, it is only prudent to establish expeditiously and unequivocally the human metabolic disposition to this substance [274, 276]. Steviol shows also some antiandrogenic activity. The authors [274] conclude that in order to display mutagenic activity, the diterpenoids must have a hydroxyl group at C_{13} in addition to a double bond that joins C_{16} and C_{17}.

2.2.2.3.2 Compounds isolated from *Rubus suavissimus* S. LEE

2.2.2.3.2.1 Rubusoside

Another member of the steviol family of glycosides — rubusoside was isolated from leaves of *Rubus suavissimus* in 5.4 % yield [277]

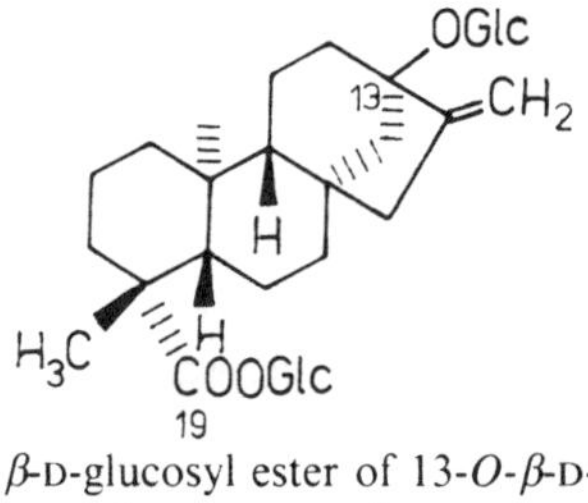

Fig. 2.16 — Rubososide — *β*-D-glucosyl ester of 13-*O*-*β*-D-glucosylsteviol. Glc — glucopyranosyl.

(Fig. 2.16). The plant is indigenous to China where tea from its leaves is a popular oriental traditional medicine. The sweet principle is a bis-glucoside of steviol, obtained from stevioside by partial enzymatic hydrolysis [278]. The authors [278] also prepared and isolated several mono- and di-bis(glucosides) by enzymic transglucosidation.

Rubusoside is 114 times sweeter than a 0.025 % solution of sucrose (when sucrose sweetness = 1, rubusoside is 150 times sweeter [270]). However, the sweetness is tainted by a slight bitter taste [226].

2.2.2.3.3 Baiyunoside and flomisoside-I

Roots of a Chinese officinal herb called Bai-Yun-Shen (*Phlomis betonicoides* DIELS, family Labiateae) contain apart from active substances used in the therapy of women's diseases, at least two additional sweet diterpenoids — baiyunoside (a) and flomisoside-I (b) (Fig. 2.17).

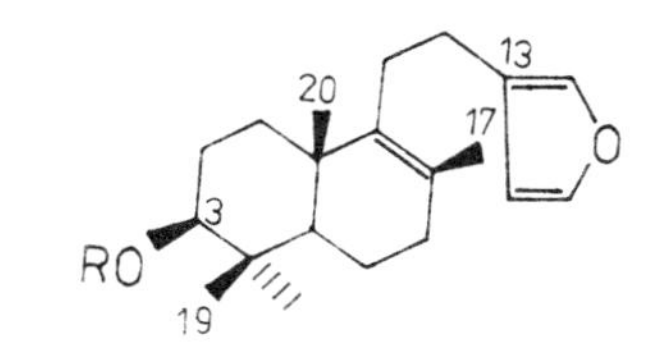

Fig. 2.17 — Structure of (a) the labdane type diterpenic glycosides baiyunoside. R = β-D-Glc-β-Xyl(2 ← 1); (b) flomisoside-I, R = β-Glc-α-Rha(2 ← 1); (c) baiyunol, R = H. Glc — glucopyranosyl, Xyl — xylopyranosyl, Rha — rhamnopyranosyl.

Dried roots are first extracted with methanol, then with butanol-1, giving very low yields (0.1 %) of the sweet glycosides [279, 280]. Subsequent enzymatic hydrolysis splits off the aglycone baiyunol which is not sweet (Fig. 2.17c). On the basis of ^{1}H NMR and mass spectra, the structures of the aglycones have been recognized as the monohydroxy diterpenes of the labdanolide family which has a β-substituted furan ring [270]. The sweetness of baiyunoside exceeds by 500 times that of sucrose and persists longer than one hour — a quality that is likely to disqualify it for use in a number of foods and drinks. Relative sweetness of flomisoside has not yet been published [280].

2.2.2.3.4 Hexahydrofluorene dicarboxylic acid

Japanese authors [281] described in 1971 the isolation of a novel type of sweet substance, structurally quite unlike any other so far known sweetener. They isolated four stereoisomers of 4β,10α-dimethyl-1,2,3,4,5,10-hexahydrofluorene-4,6-dicarboxylic acid from the resin of pine (*Pinus*); only the acid A was sweet, the other isomers were insipid (Fig. 2.18). The free acid A is 1300–1800 times sweeter; its Na salt even 1600–2000 times sweeter than a 0.0006 % solution of sucrose. Unfor-

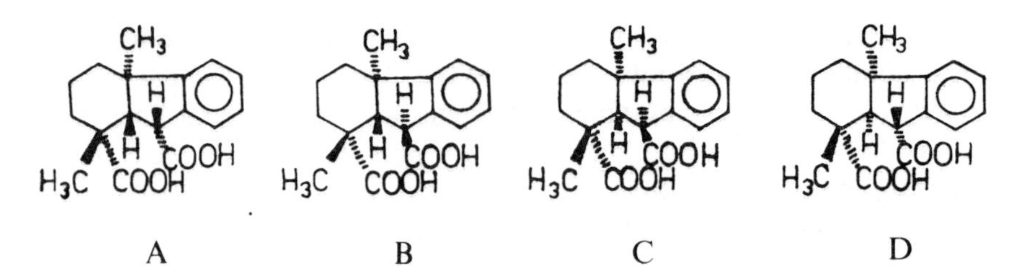

Fig. 2.18 — Stereoisomers of a hexahydrofluorene dicarboxylic acid.

tunately, the sweetness is accompanied, as in the previous case, by a significant bitter taste; it is 12–18 times more bitter than caffeine [226]. The compounds have not yet undergone a complete series of tests (carcinogenicity, toxicity and other), the possible future applications are not known either.

2.2.2.4 Triterpenoids

2.2.2.4.1 Glycyrrhizin

The root of liquorice (*Glycyrrhiza glabra* L. family Leguminosae) was a well established healing agent already in ancient medicine and pharmacy [226]. The herb has been known in the Mediterranean area for a long time, it is grown in Spain, France, Italy, as well as in the former Soviet Union (*G. uralensis* FISCH.).

Glycyrrhizin (glycyrrhizic acid), the sweet principle of the liquorice root, is a diglucuronide of the aglycone glycyrrhetinic acid, which itself is not sweet (Fig. 2.19). It can be obtained from the roots (6–14 %) and rhizomes of liquorice (23 %). Being an acid, it is usually isolated as the calcium, potassium or magnesium salt, accompanied by other components such as starch, proteins, flavanoids and resins.

Glycyrrhizin was first isolated in the pure form in 1907 [282]. The

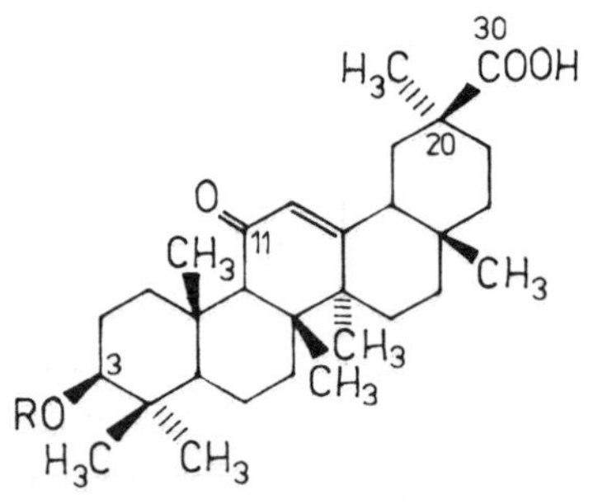

Fig. 2.19 — Structure of glycyrrhizine, R = β-GlcCOOH-β-GlcCOOH(2 ← 1) and glycyrrhetinic acid, R = H. GlcCOOH — β-D-glucuronopyranosyl.

colourless crystals are hardly soluble in cold water, but dissolve well in hot water or in ethanol. Aqueous hot solutions of the free acid and its calcium salt form gels on cooling; the acid or the salt can be crystallized as hydrates from acetic acid or ethanol. Hydrolysis of the potassium salt by dilute sulphuric acid yields crystalline glycyrrhetinic acid which is water-insoluble and without taste. A recently isolated enzyme (1988), called 'glycyrrhizinoic acid hydrolase', cleaves selectively the 3-*O*-*β*-D-glucuronide bond and thus releases the aglycone — glycyrrhetinic acid [287].

The structure of glycyrrhetinic acid was studied by several groups; it belongs to the oleane group of triterpenoidal glycosides [283–285].

The most accessible commercial derivative of glycyrrhetinic acid is the ammonium salt which is prepared by a dedicated method starting with liquid extracts from the countercurrent extraction. The salt is reasonably stable at elevated temperatures, soluble in hot or cold water, and in propylene glycol [286].

Its many applications include the use as a foaming agent, ingredient for aromatization of food, sweetening agent for the taste correction of food and drugs and in the production of confectioneries. It is also found in cigarette and pipe tobacco [286].

The ammonium salt of glycyrrhetinic acid is about hundred times as sweet as sucrose; some sources [226] give the relative sweetness as measured against 20 mg/100 ml (70× sweeter) or 10 mg/100 ml (93× sweeter) sucrose solutions. The onset of the sweet taste is rather slow, but the taste is long and is accompanied by a strange aftertaste. It shows synergism when mixed with sucrose. Attempts at improving the sensory properties of glycyrrhizin by modifying the structure of its sugar unit have so far failed [288, 289].

The structure and physiology of glycyrrhizin resemble those of a natural corticoid — 11-deoxycorticosterone [290], a fact that has stood in the way of its approval as a food additive in many countries.

Other novel sweet triterpenoidal oligoglycosides were isolated from the air-dried roots of *Glycyrrhiza punctata* BATAL (Leguminosae) (Fig. 2.20a) — apioglycyrrhizin and araboglycyrrhizin (Fig. 2.20b).

The methanolic extract of the roots yielded oligosaccharidic fractions which contain apart from the known flavanoidal glycosides, also glycyrrhizin (1.5 %), apioglycyrrhizin (0.32 %) and araboglycyrrhizin (0.14 %). Total methanolysis furnishes also the aglycone — glycyrrhetinic acid.

The structure of apioglycyrrhizin was inferred, from its physicochemical parameters and chemical reactivity, as that of 3-*O*-[*β*-D-apiofuranosyl(1→2)-*β*-D-glucuronosyl]glycyrrhetinic acid. Similarly, ara-

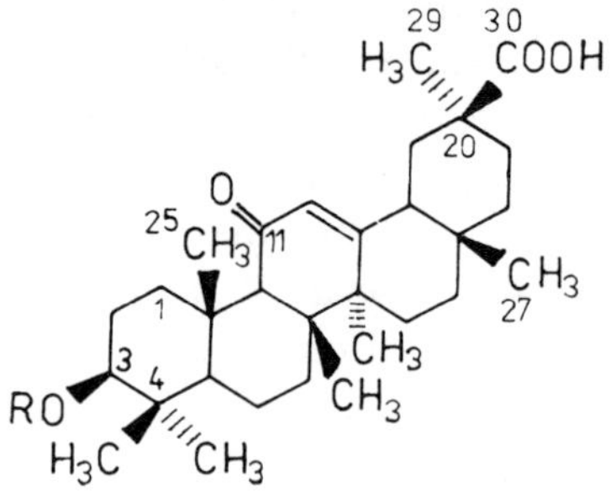

Fig. 2.20 — Structure of (a) apioglycyrrhizine, R = β-GlcCOOH-β-apioFur(2 ← 1); (b) araboglycyrrhizine, R = β-GlcCOOH-α-L-Ara(2 ← 1). β-GlcCOOH — β-D-glucuronopyranosyl, β-apioFur — β-D-apiofuranosyl, α-L-Ara — α-L-arabinopyranosyl.

boglycyrrhizin was identified as 3-*O*-[α-L-arabinopyranosyl(1→2)-β-D-glucuronosyl]glycyrrhetinic acid [291]. The investigation of their respective sweetness revealed that apioglycyrrhizin was twice as sweet, whilst araboglycyrrhizin just about equalled the sweetness of glycyrrhizin.

2.2.2.4.2 Periandrines

The roots of a sweet tasting wood *Periandra dulcis* MART., Brazilian plant from the Leguminosae family, contains 0.38 % of glycyrrhizin. A more detailed analysis showed that the sweet principle was in fact a mixture of four triterpenoidal oleane-type glycosides, designated periandrine I, II, III and IV (Fig. 2.21a, b) [292–294]. The structure of periandrine I was determined by X-ray analysis [292].

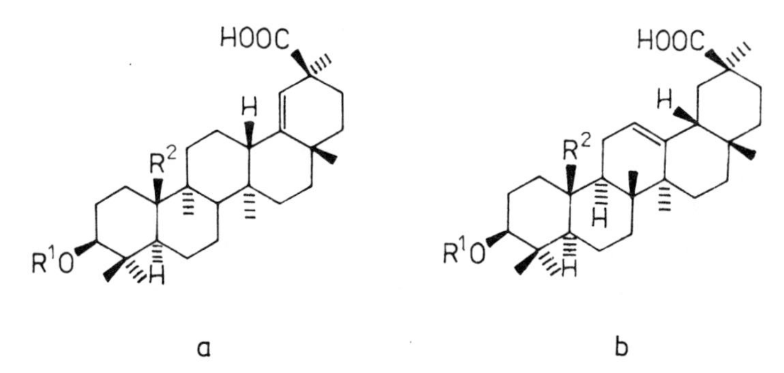

Fig. 2.21 — Periandrine I (a), R^1 = β-GlcCOOH-β-GlcCOOH(2 ← 1), R^2 = CH=O; Periandrine III (a), R^1 = β-GlcCOOH-β-GlcCOOH(2 ← 1), R^2 = CH_2OH; Periandrine II (b), R^1 = β-GlcCOOH-β-GlcCOOH(2 ← 1), R^2 = CH=O; Periandrine IV (b), R^1 = β-GlcCOOH-β-GlcCOOH(2 ← 1), R^2 = CH_2OH. β-GlcCOOH — β-D-glucuronopyranosyl.

Quantitative analysis of 30 kg of dried roots of *Periandra dulcis* yielded 300 mg of periandrine I, 170 mg of II, 23 mg of III and 9 mg of IV. The relative sweetness of pure periandrines exceeds (85–95 times) that of sucrose and about equals that of glycyrrhizin; in contrast to the latter, the sweet sensation sets in much faster.

Periandrines are not mutagenic and their acute toxicity in mice was 2 g per kg of body weight [226].

2.2.2.4.3 Mogrosides

Finely tuned high-pressure liquid chromatography (HPLC) enabled the isolation of the sweet principle of the Chinese Lo Han Kuo fruit (*Thlandiantha grosvenori* (SWINGLE) C. JEFFREY, family Cucurbitaceae) which was identified as a triterpenoidal glycoside, mogroside V (Fig. 2.22a) [295].

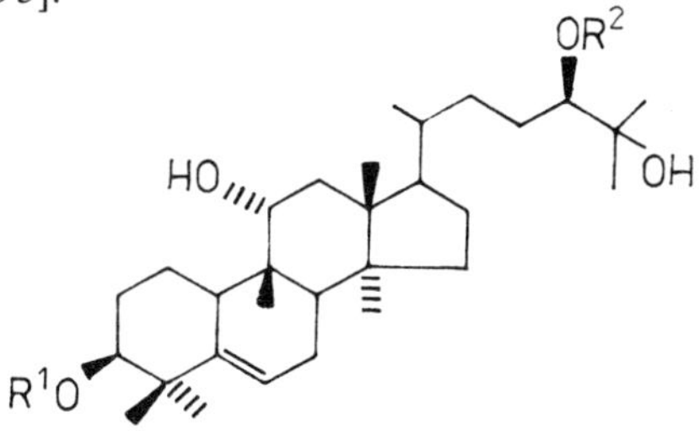

Fig. 2.22 — Mogroside V (a), $R^1 = \beta$-Glc-β-Glc($6 \leftarrow 1$), $R^2 = \beta$-Glc-β-Glc($6 \leftarrow 1$)-β-Glc($2 \leftarrow 1$) and mogrol (b), $R^1 = R^2 =$ H, β-Glc — β-D-glucopyranosyl.

The dark brown, pumpkin-shaped fruit (6–11 cm long and 3–4 cm in diameter) has a very thick skin (0.5–0.8 mm). When dried, it is used in the traditional Chinese medicine for treatment of colds, sore throats, as well as stomach and intestinal disorders.

The sweet principle from the fruit of *T. grosvenori* was first isolated and purified by Lee [296]. Takemoto and co-workers [297] have elucidated the structures of three sweet triterpenoidal glycosides from the same plant, mogroside V, the sweetest and the most abundant compound, mogroside IV and VI. The structure of mogrol, the aglycone, is shown in Fig. 2.22b. Mogroside V is water-soluble and the solutions can be heated to boiling point without decomposition. Dried fruit contains 0.81–1.29 % of mogroside V.

Pure mogrosides have a much more pleasant sweet taste than the raw extracts. At a concentration of 20 mg/100 ml, mogrosides V and VI are respectively 256 and 125 times sweeter than sucrose.

Mogroside V was shown to be non-mutagenic and produced no mortality in acute toxicity experiments on mice at doses up to 2 g/kg of body weight [295].

2.2.2.4.4 Other sweet substances isolated from plants of the Cucurbitaceae family

Roots of the plant *Bryonia dioica* JACQ. yielded a sweet matter of as yet unknown structure named bruyodulcoside [226].

Novel sweet cucurbitane-type glycosides were isolated from the Chinese officinal plant *Hemsleya carnosiflora,* designated as carnosifloside V and VI [298] (Fig. 2.23). Quantitative data on their sweetness have not yet been published. Carnosifloside I was found insipid, carnosiflosides II, III and IV were found to be bitter.

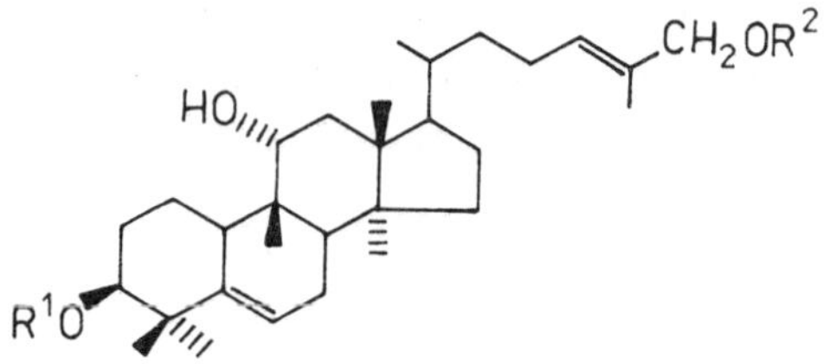

Fig. 2.23 — Carnosifloside V (a) and carnosifloside VI (b); a — R^1 = β-Glc, R^2 = β-Glc-β-Glc(2 ← 1); b — R^1 = β-Glc, R^2 = β-Glc-β-Glc(6 ← 1); Glc — D-glucopyranosyl.

2.2.3 Steroidal saponins

The extraction of rhizomes of the sweet plant *Polypodium vulgare* L. (Polypodiaceae) gave in very low yield (0.03 %) a steroidal saponin, osladin, a bis(glycoside) [299]. Its disaccharidic unit is that of 2-*O*-α-L-rhamnopyranosyl-β-D-glucopyranose.

Ferns as well as mosses are widespread in the forests of Europe, Asia and America. Both prefer shady, moist localities from sea level to about 1800 m in altitude. Sweet taste and the medicinal effects of rhizomes of Polypodiaceae plants have been known for a long time [300].

The structure of osladin, one of its sweet principles, was determined by Herout and his group [299, 301]. The configuration of the L-rhamnopyranosyl group in position C_{26} of osladin was not known at the time of its isolation [301]. The stereochemistry of the aglycone of osladin was deduced in the course of its synthesis carried out by Havel and Cerny [302]. The information on the sweetness of osladin is far from equivocal; some reports differ by an order of magnitude. Thus Kinghorn [226] estimated its value to be 3000 times that of sucrose, whilst others [303] rate it to be only 300 times sweeter. The structure of the saponin was finally determined by Kinghorn's group [304] as that of 26-*O*-α-L-rhamnopyranosylpolypodogenin-3-*O*-α-L-rhamnopyranosyl-(1→2)-β-D-glucopyranoside. Kinghorn also gave the identified saponin a new name — polypodoside A (Fig. 2.24).

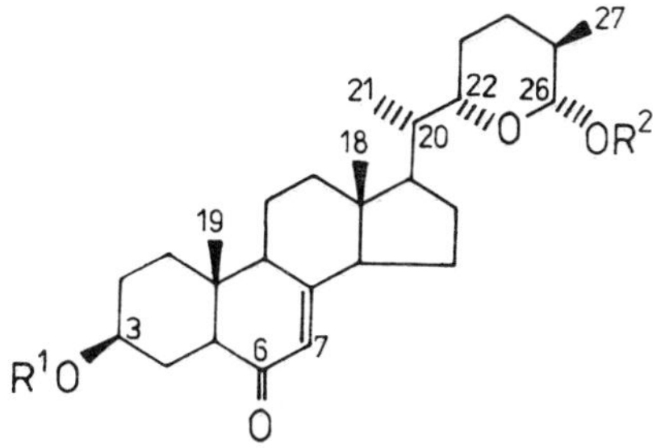

Fig. 2.24 — Polypodoside A, R^1 = β-Glc-α-Rha(2 ← 1); R^2 = α-Rha. β-Glc — β-D-glucopyranosyl, α-Rha — α-L-rhamnopyranosyl.

A tasting panel evaluated the sweetness of polypodoside A as being 600 times greater than that of a 6 % w/v sucrose solution.

Kinghorn *et al.* [304] isolated polypodoside A from the air-dried rhizomes of the North American fern *Polypodium glycyrrhiza* DC., Eaton (Polypodiaceae) in 0.29 % yield. The authors confirmed that polypodoside A was non-mutagenic (*Salmonella typhimurium* strain TM 677) and was not acutely toxic to mice.

Two novel steroidal glycosides, polypodosides B and C, were described by Kim and Kinghorn [305]. *Polypodium* rhizomes were first extracted with methanol and then with butanol. The solid residue was then separated into its constituents, polypodoside B (0.02 %) and C (0.06 %), by gel chromatography. The separation of polypodosides A–C by overpressured layer chromatography, as well as the purification of polypodoside A, was described by Fullas *et al.* [254]. Polypodoside B is only marginally sweet, polypodoside C is devoid of this effect.

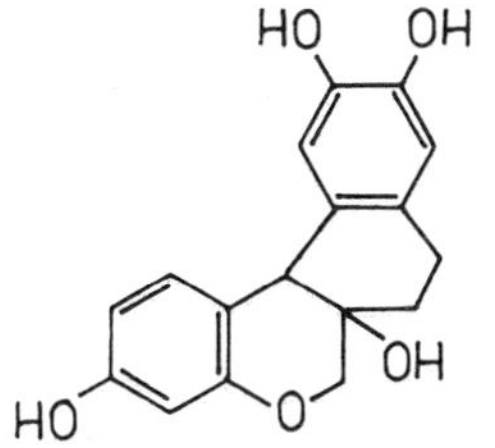

Fig. 2.25 — Brazilin.

Brazilin, another steroidal saponin (Fig. 2.25), was discovered in the Brazilian plant *Caesalpina echinata*; it was found to be 100 times sweeter than sucrose [276].

2.2.4 1*H*-2-Benzopyran-1-ones — Dihydroisocoumarins

2.2.4.1 *Phyllodulcin*

Phyllodulcin is a natural dihydroisocoumarin derivative, first isolated in 1916 from the leaves of 'amacha' plant *Hydrangea macrophylla* SERINGE, var. *thunbergii* (SIEBOLD) MAKINO (Saxifragaceae), as well as from other Hydrangeae family members which are used in Japan for the preparation of sweet teas that are required at certain rituals [306].

Asahina and Asano [307] isolated it and studied its structure, they pointed out the presence of a dihydroisocoumarin skeleton in the molecule. Absolute configuration at the asymmetric C_3-(3*R*) centre was determined in 1959 [306] (Fig. 2.26).

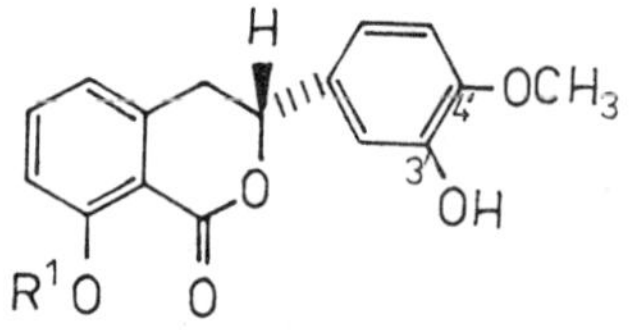

Fig. 2.26 — Phyllodulcin, R^1 = H and phyllodulcin 8-*O*-β-D-glucoside, R^1 = β-D-glucopyranosyl.

Phyllodulcin is also accessible by enzymatic hydrolysis of phyllodulcin glycoside [308].

Other Japanese authors [309] accessed phyllodulcin biosynthetically, *via* a C_6–C_3–C_6 precursor derived from shikimic acid and an acetate. Cell cultures from 'amacha' are also capable of producing phyllodulcin [308].

Chemical syntheses of phyllodulcin start either from *trans*-3-(3-benzyloxy-4-methoxyphenyl)-2-propenal [310] or from [2-(3-hydroxy-2-carboxyphenyl)]-3-methoxy-4-hydroxyphenylpropenoic acid [311].

Phyllodulcin is 400 times sweeter than sucrose and is rather popular in Japan. Nevertheless, it has several drawbacks as a sweetener — its water solubility is low, the sweet taste of dilute solutions sets in very slowly and is blemished by a persistent aftertaste [226]. Phyllodulcin has no acute toxicity in doses up to 2 g/kg of body weight of mice and is not mutagenic [226].

2.2.4.2 *Analogues of phyllodulcin*

The investigation of the structure–sweetness relationship of dihydroisocoumarins has helped to establish the following empirical rules:

(a) the presence of the 3′-hydroxy-4′-methoxyphenyl group is essential for the manifestation of sweetness
(b) heterocyclic ring need not be planar
(c) both the carbonyl C=O and the hydroxyl groups must be present; if either one is missing, sweetness decreases or even disappears [226]

Table 2.18 — Sweet compounds based on structure

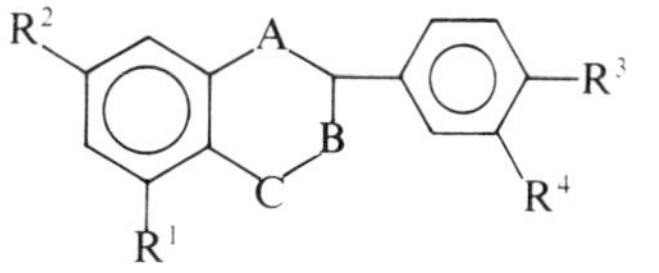

R^1	R^2	R^3	R^4	A	B	C	Rel. sweetness[a]	Ref.
OH	H	OCH_3	OH	CH_2	O	C=O	400	[313]
H	H	OCH_3	OH	CH_2	O	C=O	100	[270, 313]
OH	H	OCH_3	OH	CH_2	NH	C=O	1–10	[270]
H	H	OCH_3	OH	CH_2	NH	C=O	0	[270, 313]
H	H	OCH_3	OH	CH_2	O	CH_2	350	[270, 313]
H	H	OCH_3	OH	O	CH_2	CH_2	350	[270]
H	OH	OCH_3	OH	O	CH_2	CH_2	350	[270, 312]
H	H	OCH_3	OH	O	O	CH_2	3000	[270, 312]
H	H	OCH_3	OH	O	O	C=O	100	[270, 313]
H	H	OCH_3	OH	S	O	C=O	250	[270, 313]
H	H	OCH_3	OH	O	CH_2	O	450	[316]
H	H	OCH_3	OH	O	CH_2	C=O	0	[270, 314]

a — sucrose = 1.

The synthetic phyllodulcin analogue, 2-(3-hydroxy-4-methoxyphenyl)-1,3-benzodioxane [315] is the sweetest of all synthetic derivatives. Its very high sweetness (3000 times sweeter than sucrose) is, however, marred by its instability; the molecule is slowly hydrolysed in water, so much so that aqueous solutions stored at 25 °C for 7 days loose sweetness completely [312, 315] (Table 2.18).

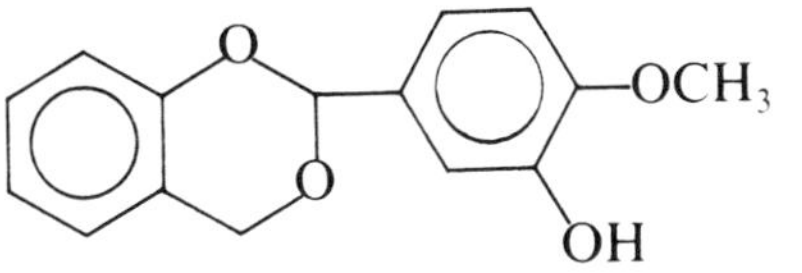

Scheme 2.16 outlines the synthesis of another phyllodulcin analogue, 2-(3-hydroxy-4-methoxyphenyl)-1,4-benzodioxane.

Using a 3 % solution of sucrose as a standard, the authors [316] estimated that its sweetness was 450 times that of sucrose. Although its

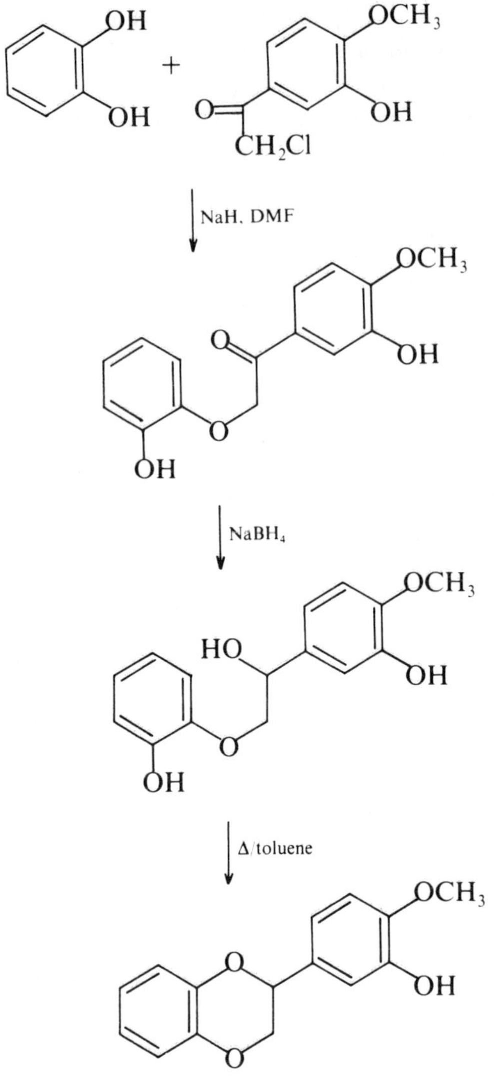
OH
OH
+
OCH3
OH
O
CH2Cl
NaH, DMF
OCH3
O
OH
O
OH
NaBH4
OCH3
HO
OH
O
OH
Δ/toluene
OCH3
O
OH
O

2-(3-hydroxy-4-methoxyphenyl)-1,4-benzodioxane
50%, m. p. 98–99° C

Scheme 2.16

water solubility is rather poor, aqueous solutions remain sweet even after one month's storage. Sweetness of the above analogue is perceived after only 2–3 seconds and it has a weak liquorice aftertaste. The authors also assume that the benzodioxane derivative has a conformation that resembles one of phyllodulcin itself, and thus interacts with the sweetness receptor in the same manner.

2.2.5 Flavanoids

Bitter or bittersweet flavanoids were first isolated from the lemon fruit more than a century ago. A major group of citrus flavanoids are flavanone-7-glycosides. They are abundantly accessible and their isolation is simple (Table 2.19).

Table 2.19 — Flavanoidal glycosides from citrus fruits

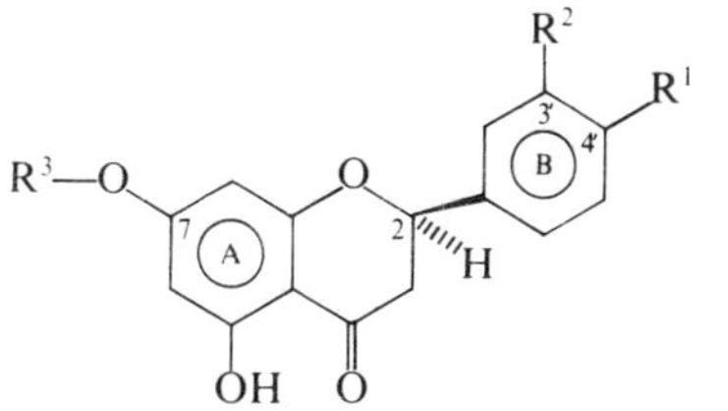

Name	R^1	R^2	R^3	Taste
Neohesperidin	OCH_3	OH	β-neohesperidosyl	bitter
Hesperidin	OCH_3	OH	β-rutinosyl	tasteless
Naringin	OH	H	β-neohesperidosyl	bitter
Naringinrutinoside	OH	H	β-rutinosyl	tasteless
Poncirin	OCH_3	H	β-neohesperidosyl	bitter
Isosakuranetinrutinoside	OCH_3	H	β-rutinosyl	tasteless
Neoeriocitrin	OH	OH	β-neohesperidosyl	bitter
Eriocitrin	OH	OH	β-rutinosyl	tasteless

Members of this group of compounds differ in their flavanone functional groups in positions 3′ and 4′, or in their respective constituent sugar (disaccharide) unit in position 7. The sugar unit in flavanoids is either β-rutinose (6-*O*-α-L-rhamnopyranosyl-β-D-glucopyranose) or β-neohesperidose (2-*O*-α-L-rhamnopyranosyl-β-D-glucopyranose).

Significantly, the two glycosidic types taste differently, the flavanone rutinosides are tasteless but the flavanone neohesperidosides are to some extent bitter.

Thus the point of attachment of rhamnose to glucose (position 2 or 6) governs the manifestation of bitterness or tastelessness. Similarly, transporting the rhamnose to the position 3 or 4 of glucose yields only slightly bitter flavanones [317, 318]. Hesperidin is a major flavanoidal component of oranges (*Citrus sinensis*) and lemons (*Citrus limon*), whilst naringin is found in grapefruit (*Citrus paradisi* MACFAD), neohesperidin in Spanish oranges (*Citrus aurantilum* L.) and poncirin in lemons (*Poncirus trifoliata*) [319].

Guadagni *et al.* [320] found that there was a wide range of sensitivity to naringin bitterness, mainly due to genetic factors. However, there is also a possibility that bitterness arises to some extent from variations in the stereochemical composition of commercial naringin.

Naringin obtained from grapefruit peel can exist in both the 2*R* and 2*S* forms. Gaffield *et al.* [321] showed that the chirality of naringin changed unexpectedly as a function of maturity of the fruit. The ratio of (2*S*)- to (2*R*)-naringin isomers isolated from immature fruit is about 9:1 but changes to about 2:3 when isolated from mature fruit. The tasters also reported that the (2*R*) isomer is more bitter than the (2*S*).

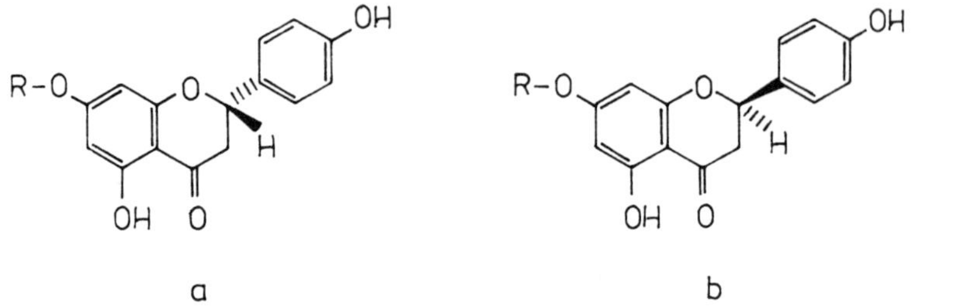

Fig. 2.27 — Absolute configuration of naringin. a – (-)-2(*S*)-naringin, b – (-)-2(*R*)-naringin. R = *β*-neohesperidosyl.

In order to learn more about the structure–sweetness relationship, many flavanone analogues have been synthesized (Table 2.20), such as 3′,5-dihydroxy-4′-methoxyflavanone (derivative 2) by condensation of 2,6-dihydroxyacetophenone with isovanillin [314] (Scheme 2.17).

Table 2.20 — Synthetic flavanoids (2,3-dihydro-2-(3-R^2,4-R^1-phenyl)-4-1*H*-benzopyran-4-ones)

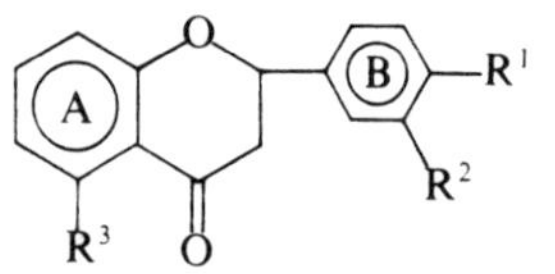

Compound	R^1	R^2	R^3	Rel. sweetness[a]	Ref.
1	OMe	OH	H	0	[270, 322]
2	OMe	OH	OH	350	[270, 314]
3	OEt	OH	OH	150	[314]
4	*O*-*n*-Pr	OH	OH	150	[314]
5	OH	OMe	OH	0	[314]
6	OH	H	OH	0	[314]

[a] Sweetness of 3% sucrose solution = 1.

The replacement of the original methoxy group by ethoxy or propoxy group decreases the sweetness; compounds that lack the hydroxyl group in position 5 or 3′ are tasteless (Table 2.20).

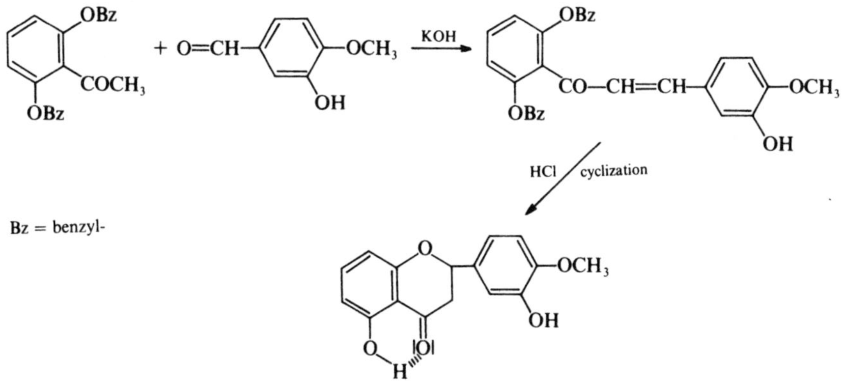

OBz
+ O=CH– –OCH_3 KOH
$COCH_3$
OBz OH
OBz
CO–CH=CH– –OCH_3
OBz OH
HCl cyclization
Bz = benzyl-
O
–OCH_3
OH
O IOI
H

Scheme 2.17

2.2.5.1 *Semisynthetic dihydrochalcones*

Semisynthetic dihydrochalcones are accessible by catalytic hydrogenation of corresponding chalcones, which are derived from naturally occurring flavanones by ring opening catalysed by a weak base, (Scheme 2.18).

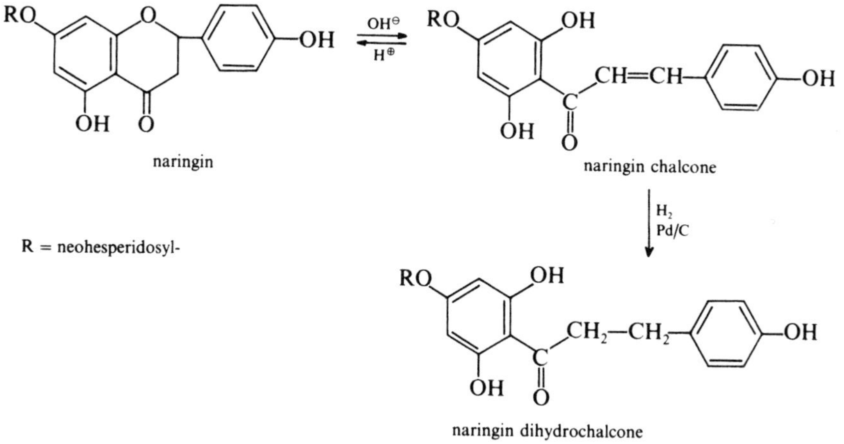

RO O –OH $OH^{\ominus}$ / $H^{\oplus}$ RO OH
CH=CH– –OH
OH O OH C
O
naringin naringin chalcone
H_2
Pd/C
R = neohesperidosyl-
RO OH
CH_2–CH_2– –OH
C
OH O
naringin dihydrochalcone

Scheme 2.18

The intense sweetness of semisynthetic dihydrochalcones (DHC), naringin and neohesperidin, was first described by Horowitz and Gentili [323, 324]. Dihydrochalcone from poncirin is bitter, one from neoeriocitrin is only slightly sweet.

Flavanones actually used in mixed sweeteners are of two basic types: β-neohesperidose containing glycosides naringin and neohesperidin, and β-rutinose containing hesperidin. The authors came to the conclusion [325, 326] that the phenolic or flavanoidal glycosides displayed bitter, sweet or bittersweet taste if their sugar component was either β-neohesperidose or β-D-glucose. On the other hand, all tested rutinosides yielded tasteless derivatives.

As shown in Scheme 2.19, a β-rutinosyl group containing hesperidin can be transformed into a sweet hesperetin dihydrochalcone 4′-β-D-glucoside by cleaving off the rhamnose.

A more complex removal of rhamnose, albeit in higher yield, can be affected enzymatically by the α-L-rhamnosidase [327].

Thus naringin dihydrochalcone (m.p. 170 °C) can be prepared in 94 % yield starting with naringin (Scheme 2.18) [323].

Alkaline hydrolysis of neohesperidine leads to an intensely bitter compound 4-acetyl-2,5-dihydroxyphenyl-1-β-neohesperidoside (below). This bitter compound undergoes a base catalysed condensation with the corresponding benzaldehyde derivative to give chalcones which on hydrogenation ($H_2/Pd/C$) furnish dihydrochalcones. As the total synthesis of

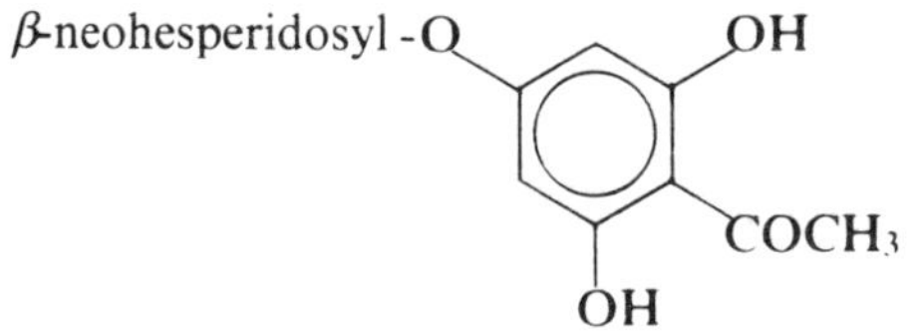

β-neohesperidosyl -O   OH
$COCH_3$
OH

dihydrochalcones is not economically viable, semisynthetic sweet chalcones are thus prepared by chemical modification of the known readily accessible compounds. For example, neohesperidin dihydrochalcones are being commercially produced from naringin which is accessible from grapefruit peel. The isolated naringin is modified by K_2CO_3 hydrolysis, then condensed with 3-hydroxy-4-methoxybenzaldehyde and finally hydrogenated.

The resulting neohesperidin DHC (82 % yield after recrystallization from methanol) has a melting point of 172–174 °C [328, 329].

Once again, many DHC analogues were synthesized for the purpose of the structure–sweetness relationship studies (Table 2.21).

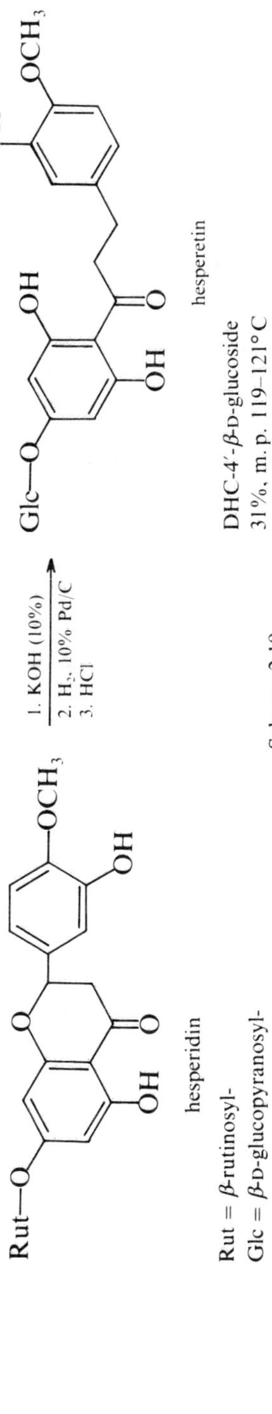
Rut—O
OH
O
OCH3
OH
hesperidin
Rut = β-rutinosyl-
Glc = β-D-glucopyranosyl-
1. KOH (10%)
2. H2, 10% Pd/C
3. HCl
Glc—O
OH
OH
O
OH
OCH3
DHC-4'-β-D-glucoside
31%, m. p. 119–121°C

Scheme 2.19

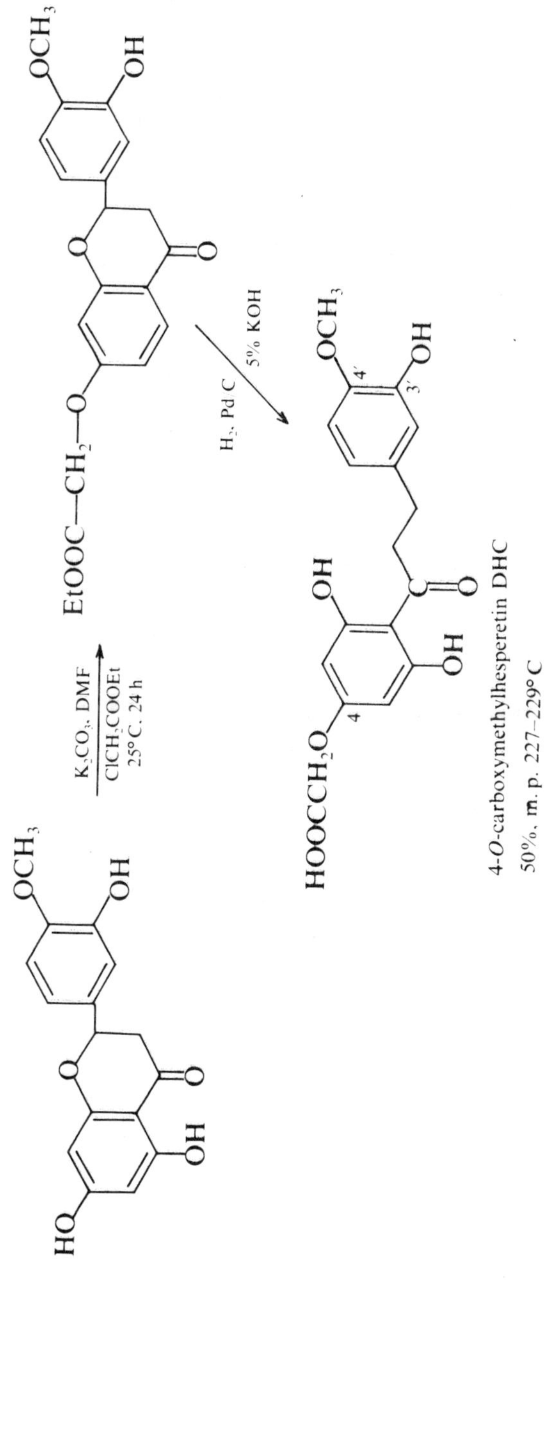
HO
OH
O
OCH3
OH
hesperetin
K2CO3, DMF
ClCH2COOEt
25°C, 24 h
EtOOC—CH2—O
O
O
OCH3
OH
H2, Pd C
5% KOH
HOOCCH2O
4
OH
OH
C
O
4'
3'
OCH3
OH
4-O-carboxymethylhesperetin DHC
50%, m. p. 227–229°C

Scheme 2.20

Table 2.21 — Dihydrochalcone derivatives and their sweetness relative to sucrose

No.	R^1	R^2	R^3	R^4	R^5	Rel. sweetness	Ref.
1	H	OH	OH	OH	β-neohesperidosyl	100–300[a]	[324, 336]
2	OH	OMe	OH	OH	β-neohesperidosyl	1000–1500 2000[b]	[324, 336]
3	OH	H	OH	OH	β-neohesperidosyl	100	[329, 336]
4	OH	OEt	OH	OH	β-neohesperidosyl	1000	[329, 336]
5	OH	*O-n*-Pr	OH	OH	β-neohesperidosyl	2000	[329, 336]
6	OH	*O-i*-Pr	OH	OH	β-neohesperidosyl	1000	[324, 336]
7	OH	OMe	H	OH	H	100	[314]
8	OH	OEt	H	OH	H	1–10	[314]
9	OH	OMe	OH	OH	H	0	[314]
10	OH	OMe	OH	OH	$-OCH_2COOK$	200	[334, 336]
11	OH	OMe	OH	OH	$-OCH_2COONa$	500	[334, 336]
12	OH	OMe	OH	OH	$-OCH(CH_3)COOH$	30	[334, 336]
13	OH	OMe	OH	OH	$-OCH(CH_3)COONa$	115	[334, 336]
14	OH	OMe	OH	OH	$-O(CH_2)_3COOK$	300	[334, 336]
15	OH	OMe	OH	OH	$-OCH_2SO_3Na$	430	[333]
16	OH	OMe	OH	OH	$-O(CH_2)_3SO_3K$	700	[333]
17	OH	OMe	OH	OH	$-O(CH_2)_3SO_3Na$	500	[333]
18	OH	OCH_2COOH	OH	OH	β-neohesperidosyl	0	[317, 336]
19	OH	OMe	OH	OH	OCH_2COOMe	0	[317, 336]
20	OH	OMe	H	OH	$O(CH_2)_2CH(NH_2)COOH$	400	[270, 335]
21	OH	OMe	OH	OH	*O*-β-D-glucosyl	80	[324, 336, 338]
22	OH	OMe	OH	OH	*O*-β-D-xylosyl	100	[324, 336, 339]
23	OH	OMe	OH	OH	*O*-β-D-galactosyl	140	[336, 340]

a — naringin DHC, b — neohesperidin DHC.

The structural modifications were first concentrated on ring B of neohesperidin DHC. The pattern of variations in this region led to the conclusion that the presence of 3′-OH and 4′-OH groups was essential for sweetness even though not the sole condition. When expressed the other way round, the absence of hydroxyl groups resulted in the loss of sweetness or even in (sweet)bitterness.

Amongst the dihydrochalcones listed in Table 2.20, all of which were prepared from synthetic flavanones, only 2,3′-dihydroxy-4′-methoxy-dihydrochalcone was 100 times sweeter than a 3 % solution of sucrose; other compounds were tasteless (Table 2.21, derivatives 7–9).

Dihydrochalcones that contained carboxymethyl, sulphoalkyl or homoseryl group were also prepared with an aim of increasing the water solubility (Table 2.21).

A simple and widely applicable method for the preparation of 4-*O*-carboxyalkyl analogues was devised [330, 331] (Scheme 2.20).

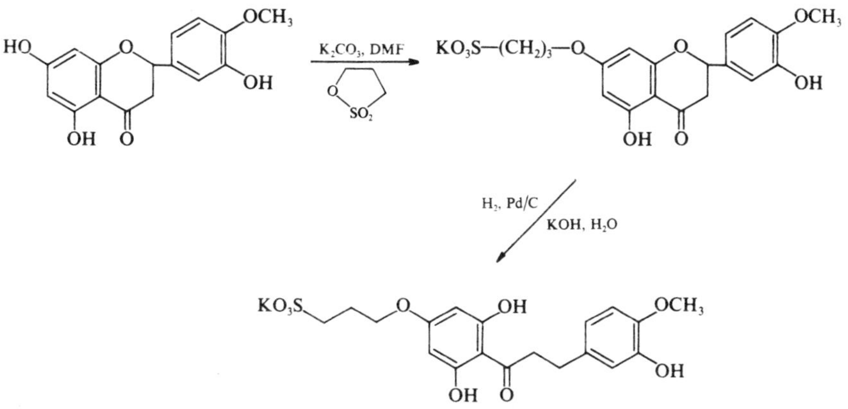

Scheme 2.21

Hesperetin was obtained by acid hydrolysis (H_2SO_4/CH_3OH) of hesperidin [332].

The synthesis of 4-*O*-sulphopropylhesperetin DHC utilizes the regioselective alkylation of hesperetin with 1,3-propane sultone, followed by hydrogenation in alkaline media [333] (Scheme 2.21).

The carboxymethyl group must be located at the position 4′ in order to retain sweetness; when it occupies any other position, sweetness is lost. Esterification of the 4-carboxymethyl group results in the loss of sweetness also.

Another class of DHC analogues, that contained other sugar residues such as glycosyl, galactosyl and xylosyl, was also prepared (Table 2.21, derivatives 21–23).

By analogy with other natural glycosides (stevioside, glycyrrhizin) that have two glycosidic groups, Japanese authors [341] prepared several DHC bis(glycosides). Their range of sweetness ranged from 400 times to only slight sweetness, tastelessness or even bitterness.

L-Rhamnose unit can be replaced by other L-sugars without the loss of sweetness, its replacement by D-sugars suppresses sweetness. Sweetness prevails if one or both monosaccharide units are substituted.

Figure 2.28 depicts hesperidin DHC analogues [331, 342]. None of the synthetic amides were sweet; the N–CH_3 derivative was bitter [342].

Wong and Horowitz [343] were able to determine the structure of neohesperidin DHC by single crystal X-ray analysis. They found both

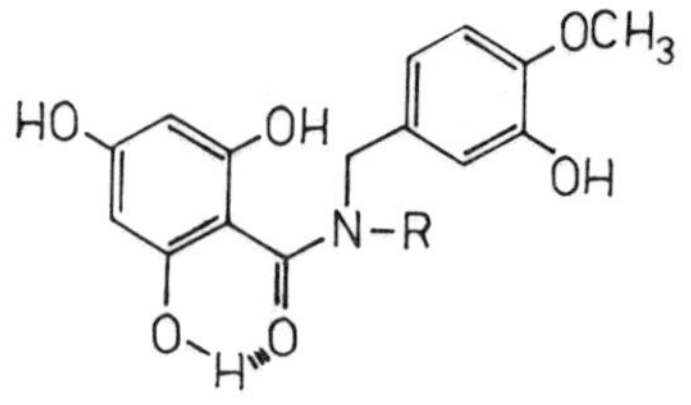

R = H, Me

Fig. 2.28 — Dihydrochalcone amides.

the rhamnosyl and glucosyl units in chair conformations. Furthermore, the aglycone unit was found to have a flat W pattern. The crystal structure consists of two (crystallographically) independent molecules which differ only in the angle of torsion of ring B.

Figure 2.29 shows the superposition of the two independent molecules.

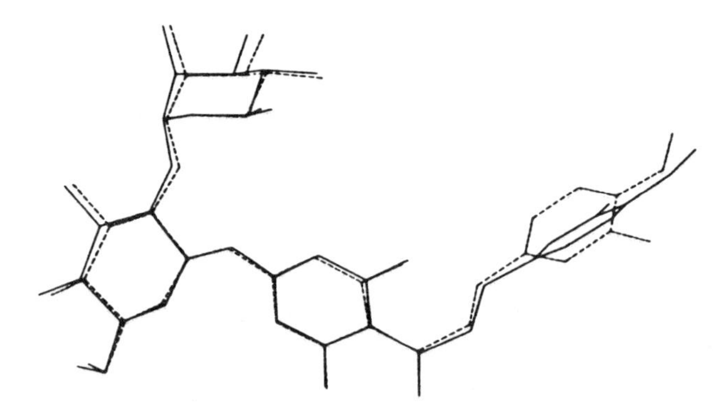

Fig. 2.29 — Superposition of two crystallographically independent molecules, broken lines indicate conformer A.

Dihydrochalcones are only sparingly soluble in distilled water at room temperature (naringin DHC 0.94 g/l, neohesperidin DHC 0.5 g/l and hesperetin DHC 4′-β-D-glucoside 0.3 g/l). The solubility of sodium and calcium salts of neohesperidin DHC is lower than 1 g/ml [344]. The aqueous solutions (down to pH 2) are quite stable at room temperature towards the hydrolysis of the saccharide component. Buffered solutions are stable for 8 hours at 10 °C in the pH range of 2–10 [225]. Sodium and calcium salts are soluble in dilute alkali, methanol and higher alcohols.

Quantitative assay of dihydrochalcones in food and drinks is carried out at present by HPLC or thin-layer chromatography (TLC)

[337, 345]. The major factor that limits the use of naringin DHC as a sweetener is its poor solubility in water: on the other hand, the taste quality limits the application range of both neohesperidin and naringin DHCs. The structure–sweetness relationship studies by some authors [333, 346] were interpreted in terms of two AH–B units being present in the DHC sweeteners (Fig. 2.30).

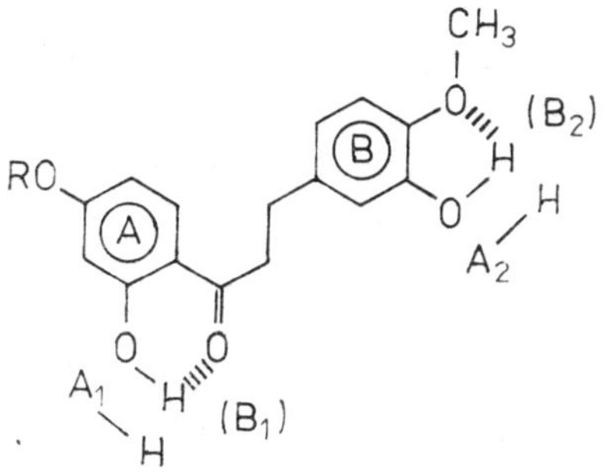

Fig. 2.30 — Structural elements of 4-*O*-substituted DHC, responsible for the sweet taste.

Semisynthetic dihydrochalcones are characterized by a menthol-like refreshing sweet taste. The sweet sensation persists longer (10 minutes) than in sucrose. Furthermore, chalcones suppress successfully, to a differing extent, the unpleasant bitter taste in some drinks and pharmaceuticals. They are 2.5–7 times sweeter than saccharin and 25–40 times sweeter than sodium cyclamate. Neohesperidin DHC is used in much smaller doses than other sweeteners because of its very high sweetness. It can be used either alone or in conjunction with other sweeteners.

Dihydrochalcone sweeteners have found application in foods, soft drinks, confectioneries and drugs. Their long-lasting sweetness and stability qualifies them for the use in chewing gums, tooth-pastes, mouthwashes and similar products. There are many examples of such applications in the patent literature, both in pure form and in admixture with sucrose or artificial sweeteners [225, 317].

Semisynthetic dihydrochalcones showed no toxicity in mice and dogs in doses of 0.2 to 1 g/kg of body weight; all measured biochemical parameters remained at normal levels. They are also non-mutagenic, non-carcinogenic and non-cariogenic. The DHC sweeteners metabolize fast in the human body; the major metabolite found in urea was 3-phenylpropanoic acid, produced by bacterial enzymes in the intestines. The metabolism of neohesperidin DHC was investigated by using a ^{14}C

labelled material [225]. Because of their low toxicity and high sweet potency, dihydrochalcones have been accepted as sweeteners and have received official clearance in several countries.

2.2.5.2 Dihydroflavanols — 2,3-dihydro-2-(3-hydroxy-4-methoxyphenyl)-3,5,7-trihydroxy-4H-1-benzopyran-4-ones

A novel sweet compound, dihydroquercetin-3-acetate (400 times sweeter than sucrose) has recently been isolated from the buds of the Paraguayan medicinal herb *Tessaria dodoneifolia* (Asteraceae) [346] (Fig. 2.31a). An analogous artificial sweetener was synthesized from the commercially available 2,4,6-trihydroxyacetophenone and isovanillin, dihydroquercetin-3-acetate-4′-methyl ether (Fig. 2.31b); its sweetness was about the same as that of the natural product. It elicits a pure, no bitter-sweet sensation, albeit rather slowly [227]. Japanese workers [347] were the first to recognize the sweetness of dihydroflavanol glycosides. They were able to isolate a sweet compound, dihydroflavanol 3-*O*-α-L-rhamnosyl(2*S*,3*S*)taxifoline, in 0.01 % yield from the dried leaves of a popular Chinese medicinal herb huang-qi (*Engelhardtia chrosolepis*) (Fig. 2.31c). Its relative sweetness or the results of the toxicological tests have not been published so far.

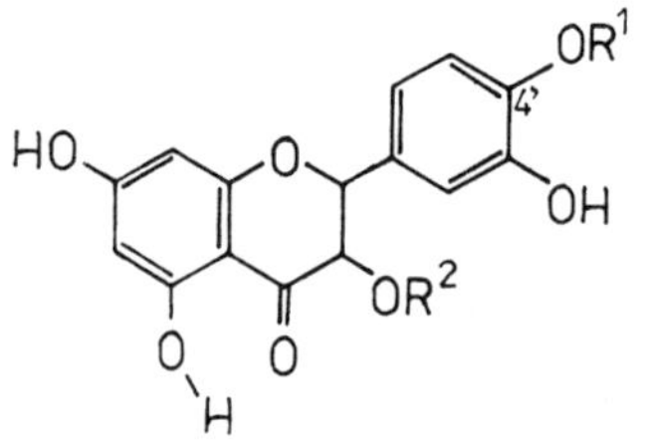

Fig. 2.31 — Sweet dihydroflavonols.

	R^1	R^2	Compound
a	H	CH_3CO	dihydroquercetin-3-acetate
b	CH_3	CH_3CO	dihydroquercetin-3-acetate-4′-methylether
c	H	α-L-Rha	3-*O*-α-L-rhamnosyltaxifoline

2.3 SWEET TASTE MODIFIERS

It is known that some compounds of plant origin are capable of inducing or suppressing the sweet sensation. Clearly, such compounds are predestined for the role of food additives as taste modifiers, or as models for the study of the mechanism responsible for eliciting sweet taste at the molecular level [227].

2.3.1 Inducers of sweet taste

2.3.1.1 Miraculin

Miraculin is a taste modifying protein capable of making acids taste sweet (less sour). This property was discovered in 1852 [348] and confirmed later [214] during the reinvestigation of miraculin's unusual sensory properties. It is an active principle of the berries of the western African shrub *Richardella dulcifica* (Sapotaceae), previously known under the name *Synsepalum dulcificum* DANIELLI. The active principle could be extracted from the pulp of the red coloured, oval shaped, miracle fruit with 0.5 *M* sodium chloride solution of neutral pH. The adjective 'miracle' was aptly added after it had been found that chewing of the berries altered the taste of any sour or acid tasting drink taken subsequently, thus vinegar tastes like port wine and the sharp sour taste of lemon is reduced to that of lemon flavoured soft drink. Furthermore, the effect is long-lasting — a somewhat disturbing feature in the case of meals with mixed tastes.

Inglett *et al.* [349] were the first to investigate the isolation and the characterization of this taste modifying substance. The isolation of miraculin is a difficult task because of its low concentration in the fruit and its high pH sensitivity.

A number of methods of isolation from the miracle fruit have been devised. Kurihara and Beidler [350], and independently also Brouwer *et al.* [351], reported that the active part of miraculin was a glycoprotein. Shortly thereafter, Kurihara and Terasaki [352] were able to purify and separate three active protein components, with molecular weights in the range of 40,000–41,000, from an extract of *R. dulcifica.* Apart from the proteins, the authors also identified various amounts (16–41 %) of saccharides. Disparate sugar content values have been reported, for example 6.7 % arabinose and xylose [350], 7.5–21 % sugar content, comprising of glucose, arabinose, galactose and rhamnose [351]. On the other hand, higher purity samples were reported to contain 6.3 % of fructose, xylose, mannose and galactose [353]. The disparate sugar assay values reflect the differences in the method of isolation and purification. Miraculin is thermostable in the pH range 3–12.

The protein moiety of miraculin consists of 373 amino acids (Table 2.22) and is characterized by an isoelectric point at 8.3–9 [353, 354].

Miraculin has the distinction of being the largest known molecule able to elicit a taste sensation, and in spite of the numerous, mostly speculative, mechanistic proposals [216], its mode of action remains a

Table 2.22 — Number of amino acid residues in miraculin

Amino acid		Number of residues
Asparagine	Asp	42
Glutamic acid	Glu	24
Serine	Ser	24
Threonine	Thr	31
Proline	Pro	26
Glycine	Gly	34
Alanine	Ala	13
Valine	Val	37
Leucine	Leu	20
Isoleucine	Ile	15
Methionine	Met	2
Phenylalanine	Phe	26
Tyrosine	Tyr	13
Cysteine	Cys	—
Histidine	His	3
Lysine	Lys	22
Arginine	Arg	20
Tryptophan	Trp	9
Total		373

mystery. Kurihara and Beidler [355] suggested that an acid environment changed the shape of the molecule to such an extent as to allow the arabinose-xylose to simulate the sweet site. However, the protein backbone seems to be an important prerequisite of sweetness because proteolytic modification leads to the loss of activity.

Another postulate, proposed by Dzendolet [356], requires that miraculin blocks the sour receptor sites and allows the sweet taste to be generated by the ammonium group of an acid molecule. The above theories, as well as other proposals, were dealt with in more detail by Bartoshuk and co-workers [357] who concluded that miraculin in the presence of acid adds sweetness, reducing at the same time sourness by mixture suppression. According to Brouwer *et al.* [358] miraculin acts only on those structures in the tastebud cell membranes that are involved in the perception of sweet taste, making them sensitive to acids. None of the proposed theories can satisfactorily explain the potentiation effect. This remarkable effect is of considerable commercial interest, sufficient to warrant patent applications for the use in chewing gums, mouthwashes, etc. Two modes of application have been suggested for miraculin. One of them recommends the consumption of miraculin preparation shortly before the intake of non-sweet food, the other suggests a direct admixture to food. At a time, when the use of miraculin appeared sufficiently promising, the Massachussetts based company, the Miraculin Corporation, established large-scale plantations of *R. dulcificum* in West Indies and in Brazil, and then introduced into the market the miracle fruit concentrate (MFC) in the tablet form. Finally in May 1977, all products from *R. dulcificum* were denied the status of food additives in the United States [226].

2.3.1.2 Chlorogenic acid and cynarin

It has been known for some time that artichokes (*Cynara scolymus* L., family Compositae) possess taste-modifying properties in a similar way to miraculin. Extensive tests revealed that solutions of varying concentrations of a range of sensoric compounds, such as sucrose, citric acid, quinine hydrochloride and sodium chloride, elicited a sweet sensation of about the same intensity, provided the mouth was first rinsed out with an artichoke extract. Artichokes themselves taste sweet only after an induction period of about 4–5 minutes [359]. Two of the compounds that are responsible for the taste-modifying properties are chlorogenic acid (b) and cynarin (a) (Fig. 2.32).

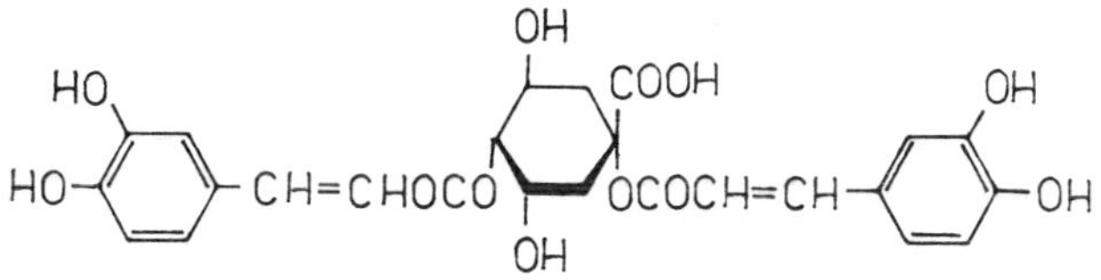

a

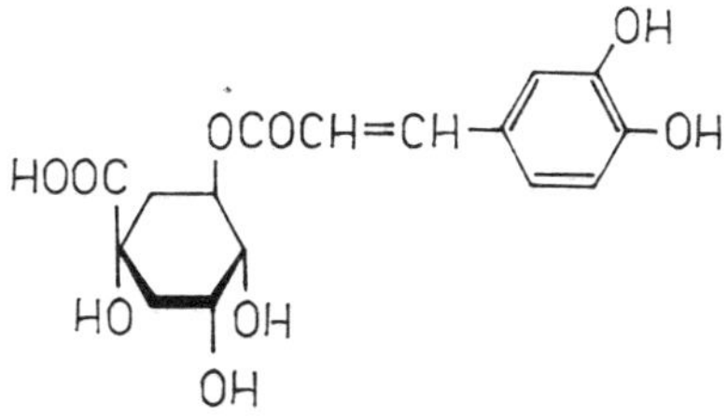

b

Fig. 2.32 — Cynarin (a) and chlorogenic acid (b).

Both compounds can be obtained by an alcohol extraction of fresh leaves of artichoke. The constitution of sweet cynarin was investigated by Panizzi and Scarpati [360]. It was obtained as a colourless crystalline compound, sparingly soluble in either cold or hot water, slightly more soluble in alcohol and acetic acid. Cynarin decomposes in alkaline media into 1,3,4,5-tetrahydroxycyclohexanecarboxylic acid and two molecules of 3,4-dihydroxycinnamic acid.

It has been suggested [359] that if it were possible to modify in a controlled manner the sweet taste of artichokes, they could be used as an ideal taste modifier for some unconventional tasteless foods, such as algae. There is not any report in the literature as yet that deals with the practical application of artichokes as a taste-modifying food additive.

2.3.2 Sweetness suppressors

The taste-modifying properties of *Gymnema sylvestre* R. Br. (Asclepiadaceae) leaves were first described more than a century ago [361]. Some compounds isolated from two plants are known to reduce selectively the sweet perception in humans. The compound responsible for the reaction is a triterpene saponin glycoside — gymnemic acid, a name given to the active component by G. S. Hooper in his paper that was published in 1887 [362].

The main components of gymnemic acid are β-D-glucuronides of the aglycone gymnemagenin that is acetylated to a varying extent (Fig. 2.33). These saponins are difficult to purify because the mixture undergoes easily transesterification reaction [227]. The structure of the aglycone gymnemagenin (Fig. 2.33) was confirmed by Stöcklin [363].

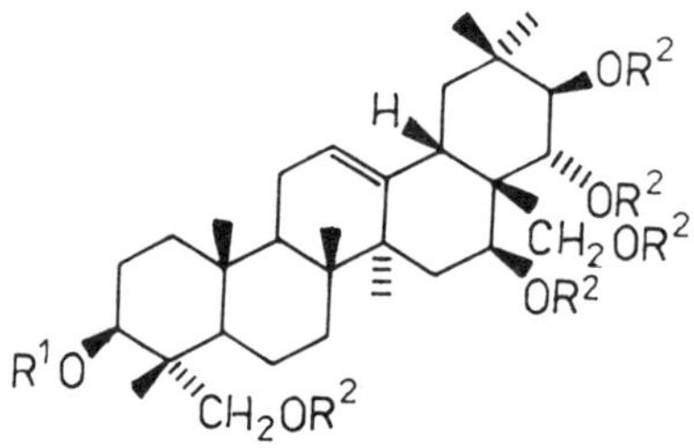

Fig. 2.33 — Structure of gymnemic acid, R^1 = β-GlcCOOH, R^2 = acyl or hydrogen, and of gymnemagenine, R^2 = H. GlcCOOH = D-glucuronopyranosyl.

The correct perception of sweet taste, once reduced by the action of gymnemic acid, does not return to normal for a period of time that can span from 20 minutes to 24 hours [364].

Another selective taste-suppressing agent was isolated from the leaves of *Hovenia dulcis* trees (family Rhamnaceae) which is indigenous to Himalaya, China, Japan and Korea. The active principle, hodulcin, is a triterpenoidal saponin, the aglycone of which, hodulcigenin (Fig. 2.34) was named and analysed by Kennedy and co-workers [364].

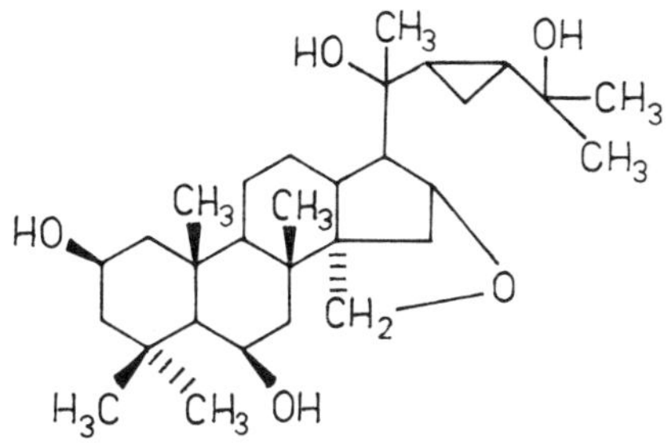

Fig. 2.34 — Hodulcigenin.

They established that *H. dulcis* saponins contain similar aglycones as those found in ziziphins (jujubogenin), and thus display similar taste-altering activity.

The selectivity of hodulcin is restricted to the reduction of sweetness only; it does not affect bitterness, saltiness or sourness. The perception of sweetness returns to normal within 1–4 minutes. Neurophysiological studies with hodulcin were carried out by Kolodny and Kennedy [365]. The understanding of the taste-altering action of plant extracts, based on systematic studies of chemical properties and the mechanism of action, could help to elucidate the mechanism of sweet perception, and thus eventually suggest novel approaches to sweetening of foods [364].

REFERENCES

1. Shallenberger, R. S. (1982) *Advanced Sugar Chemistry. Principles of Sugar Stereochemistry.* Ellis Horwood, Chichester.
2. Khan, K. (1979) In: Birch, G. G. & Parker, K. (eds) *Sugar Science and Technology.* Appl. Sci. Publ. Ltd., London, p. 181.
3. Jenner, M. R. (1980) In: Lee, C. L. (ed.) *Developments in Food Carbohydrates.* Appl. Sci. Publ. Ltd., London, p. 91.
4. Hough, L. (1985) *Chem. Soc. Rev.* **14**, 357.
5. Hassid, W. Z., Doudoroff, M. & Barker, H. A. (1944) *J. Am. Chem. Soc.* **66**, 1416.
6. Hassid, W. Z. & Doudoroff, M. (1950) *Adv. Enzymol.* **10**, 123.
7. Kraus, H. (1962) *Z. Naturforsch.* **176**, 698.
8. Lemieux, R. U. & Huber G. (1953) *J. Am. Chem. Soc.* **75**, 4118.
9. Tsuchida, H. & Komoto, M. (1963) *Agric. Biol. Chem.* **29**, 239.
10. Ness, R. K. & Fletcher, J. G. H. (1971) *Carbohydr. Res.* **17**, 465.
11. Iley, D. F. & Fraser-Reid, B. (1975) *J. Am. Chem. Soc.* **97**, 2563.
12. Beevers, C. A. & Cochran, W. (1947) *Proc. Roy. Soc. London*, Ser. A, **190**, 257.
13. Beevers, C. A., McDonald, T. R. R., Robertson, J. H. & Stern, F. (1952) *Acta Crystallogr.* **5**, 689.
14. Hough, L., Mufti, K. S. & Khan, R. (1972) *Carbohydr. Res.* **21**, 144.

15. Brown, G. M. & Levy, H. A. (1973) *Acta Crystallogr.*, Sect. 13 **29**, 790.
16. Bock, K. & Lemieux, R. U. (1982) *Carbohydr. Res.* **100**, 63.
17. Christofides, J. C. & Davies, D. B. (1985) *J. Chem. Soc. Chem. Commun.* 1533.
18. Davies, D. B. & Christofides, J. C. (1987) *Carbohydr. Res.* **163**, 269.
19. Ball, F. H., Bissett, F. H. & Chalk, R. C. (1977) *Carbohydr. Res.* **55**, 149.
20. Khan, R. (1984) *Pure Appl. Chem.* **56**, 833.
21. Hough, L. & Mufti, K. S. (1973) *Carbohydr. Res.* **29**, 291.
22. Khan, R., Mufti, K. S. & Jenner, M. R. (1973) *Carbohydr. Res.* **30**, 183.
23. Suami, T., Ikeda, T., Nishiyama, S. & Adachi, R. (1975) *Bull. Chem. Soc. Jpn.* **48**, 1953.
24. Khan, R. & Jenner, M. R. (1976) *Carbohydr. Res.* **48**, 648.
25. Percival, E. G. V. (1935) *J. Chem. Soc.* 648.
26. Bredereck, H., Hagelloch, G. & Hambsch, E. (1952) *Chem. Ber.* **87**, 1.
27. Lindley, M. G., Birch, G. G. & Khan, R. (1975) *Carbohydr. Res.* **43**, 360.
28. Lindley, M. G., Birch, G. G. & Khan, R. (1976) *J. Sci. Food Agric.* **27**, 140.
29. Hough, L. & Khan, R. (1978) *Trends Biochem. Sci.* **3**, 61.
30. Brit. 1543167 and 1543168 (1979) [*Chem. Abstr.* **91**, 1979, 193577].
31. Jenkins, J. D. & Thang, S. (1984) *Aust. J. Chem.* **37**, 1925.
32. Lee, C. K. (1987) *Carbohydr. Res.* **162**, 53.
33. Helferich, B. (1921) *Dtsch. Ber. Chem. Ges.* **54**, 1082.
34. Cottrell, A. G., Buncel, E. & Jones, J. K. N. (1966) *Can. J. Chem.* **44**, 1483.
35. Jones, J. K. N., Perry, M. B. & Turner, J. C. (1960) *Can. J. Chem.* **38**, 1122.
36. Bragg, P. D., Jones, J. K. & Turner, J. C. (1959) *Can. J. Chem.* **37**, 1412.
37. Ballard, J. M., Hough, L., Richardson, A. C. & Fairclough, P. H. (1973) *J. Chem. Soc., Perkin Trans. I* 1524.
38. Hough, L., Phadnis, S. P. & Tarelli, E. (1975) *Carbohydr. Res.* **44**, 37.
39. Parolis, H. (1976) *Carbohydr. Res.* **48**, 132.
40. Anisuzzaman, A. K. M. & Whistler, R. M. (1980) *Carbohydr. Res.* **78**, 185.
41. Hough, L. & Phadnis, S. P. (1976) *Nature* **263**, 800.
42. Birch, G. G. (1977) In: Le Magnen, J. & MacLeod, P. (eds) *Olfaction and Taste VI*. Information Retrieval, London, p. 27.
43. Birch, G. G. & Lee, C. K. (1976) *J. Food Sci.* **41**, 1403.
44. Konenko, O. K. & Kestenbaum, I. L. (1961) *J. Appl. Chem.* **11**, 7.
45. Clode, D. M., Laurie, N. A., McHale, D. & Sheridan, J. B. (1985) *Carbohydr. Res.* **139**, 147, 161.
46. Clode, D. M., McHale, D., Sheridan, J. B., Birch, G. G. & Rathbone, E. B. (1985) *Carbohydr. Res.* **139**, 141.
47. Guthrie, R. D., Jenkins, I. D., Thang, S. & Yamasaki, R. (1983) *Carbohydr. Res.* **121**, 109.
48. Eur. Pat. Appl. 67535 (1982) [*Chem. Abstr.* **98**, 1983, 161107].
49. Khan, R. & Mufti, K. S. (1975) *Carbohydr. Res.* **43**, 247.
50. Otake, T. (1974) *Bull. Soc. Chem. Jpn.* **47**, 1938.
51. Binkley, W. W., Horton, D. & Bhacca, N. S. (1969) *Carbohydr. Res.* **10**, 245.
52. Lemieux, R. U. & Nagarajan, R. (1964) *Can. J. Chem. Soc.* **42**, 1270.
53. Allerhand, A. & Dodell, D. (1971) *J. Am. Chem. Soc.* **93**, 2777.
54. Ho-Sze-Cheung, Koch, H. J. & Stuart, R. S. (1978) *Carbohydr. Res.* **64**, 251.
55. Hough, L., Phadnis, S. P., Taralli, E. & Price, R. (1976) *Carbohydr. Res.* **47**, 151.
56. Almquist, R. C. & Reist, E. J. (1974) *J. Carbohydr. Nucleosides* **1**, 461.
57. Almquist, R. G. & Reist, E. J. (1976) *J. Carbohydr. Res.* **46**, 33.
58. Ball, D. H., Jenner, M. R. & Mufti, K. S. (1975) *Carbohydr. Res.* **39**, 253.

59. Kochetkov, N. R., Molodtsov, N. V. & Chizhov, O. S. (1968) *Tetrahedron* **24**, 5587.
60. Binkley, W. W., Dougherty, R. C., Horton, D. & Wander, J. D. (1971) *Carbohydr. Res.* **17**, 121.
61. Biemann, K., Dejongh, D. D. & Schnoes, H. K.(1963) *J. Am. Chem. Soc.* **85**, 1763.
62. Brucke, C. (1983) *Trends in Biotechnol.* **1**, 67.
63. MacDonald, I. & Braithwaite, A. M. (1964) *Clin. Sci.* **27**, 23.
64. Reiser, S., Hallfrisch, J., Michaelis, D. E., Lazar, F. L., Martin, E. & Prather, E. S. (1979) *Am. Chem. Clin. Nutr.* **32**, 1659.
65. Palumbo, P. J., Briones, E. R., Nelson, R. A. & Kottke, B. A. (1977) *Am. J. Clin. Nutr.* **30**, 394.
66. Liu, G., Coulston, A., Hollenbeck, C. & Reaven, G. (1984) *J. Clin. Endocrinol. Metab.* **59**, 636.
67. Little, J. A., Birchwood, D. S., Simmons, M. A., Antar, A. K., Buckley, G. C. & Csima, A. (1970) *Atherosclerosis* **11**, 173.
68. Nestel, P. J., Carroll, K. P. & Havenstein, N. (1970) *Metabolism* **19**, 1.
69. Porikos, K. P. & van Italie, T. B. (1983) *Am. J. Med.* **75**, 624.
70. Crapo, P. A., Reaven, G., Olefsky, J. & Alto, P. (1976) *Diabetes* **25**, 741.
71. Collings, P., Williams, C. & MacDonald, I. (1981) *Br. Med. J.* **282**, 1032.
72. Jenkins, D. J. A., Wolever, T. M. S., Taylor, R. H., Fielden, H., Baldwin, J. M., Bowling, A. C., Newman, H. C., Jenkins, A. L. & Goff, D. V. (1981) *Am. J. Clin. Nutr.* **34**, 362.
73. Mann, J. I. & Truswell, A. S. (1972) *Br. J. Nutr.* **27**, 395.
74. Cooper, P. L., Wahlqvist, M. L. & Simpson, R. W. (1986) *Recent Adv. Clin. Nutr.* **2**, 271.
75. Bretschneider, R. (1980) *Sucrose Technology*, 2nd ed. SNTL — Alfa, Prague (in Czech).
76. Pohlova, M. & Kastl, F. (1979) *Sweeteners in Food Industry, Fructose Syrups.* Research Institute of Food Industry, Prague, p. 7 (in Czech).
77. Tanret, C. (1895) *Compt. Rend.* **120**, 1001.
78. Tanret, C. (1894) *Bull. Soc. Chim. Fr.* **13**, 728.
79. Isbell, H. S. & Pigman, W. W. (1937) *J. Org. Chem.* **1**, 505.
80. Richards, E. M., Faulkner, J. J. & Lowry, T. M. (1927) *J. Chem. Soc.* 1733.
81. Riiber, C. N. & Minsaas, J. (1927) *Chem. Ber.* **60**, 2402.
82. Hudson, C. S. & Brauns, D. H. (1916) *J. Am. Chem. Soc.* **38**, 1222.
83. Shallenberger, R. S. (1978) *Pure Appl. Chem.* **50**, 1409.
84. Shallenberger, R. S. (1979) *Zuckerindustrie* **104**, 121.
85. Lindley, M. G. & Birch, G. G. (1975) *J. Sci. Food Agric.* **26**, 117.
86. Martin, O. R., Tomola, S. K. & Szarek, W. A. (1982) *Can. J. Chem.* **60**, 1857.
87. Shallenberger, R. S. (1963) *J. Food Sci.* **28**, 284.
88. Shallenberger, R. S. & Acree, T. E. (1967) *Nature* **216**, 480.
89. Shallenberger, R. S. & Acree, T. E. (1969) *J. Agric. Food Chem.* **17**, 701.
90. Lemieux, R. V. & Brewer, J. T. (1973) In: Isbell, H. S. (ed.), *Carbohydrate in Solutions.* Adv. Chem. Ser. 117. Am. Chem. Soc., Washington, D. C., p. 121.
91. Krier, L. B. (1972) *J. Pharm. Sci.* **61**, 1394.
92. Takagi, S. & Jeffrey, G. A. (1977) *Acta Crystallogr. B* **33**, 3510.
93. Szarek, W. A., Korppi-Tommola, S.-L., Martin, O. R. & Smith, V. H., Jr. (1984) *Can. J. Chem.* **62**, 1506.
94. Szarek, W. A., Korppi-Tommola, S.-L., Shurvell, H. F., Smith, V. H., Jr. & Martin, O. R. (1984) *Can. J. Chem.* **62**, 1512.
95. Suami, T., Ogawa, T. M., Takata, M., Yasuda, K., Suga, A., Takei, K. & Uematsu,

Y. (1985) *Chem. Lett.* 719.
96. Suami, T., Ogawa, S., Takata, M., Yasuda, K., Takei, K. & Suga, A. (1986) *Bull. Chem. Soc. Jpn.* **59**, 819.
97. Suami, T., Ogawa, S., Uematsu, Y. & Suga, A. (1986) *Bull. Chem. Soc. Jpn.* **59**, 1261.
98. Ogawa, S., Uematsu, Y., Yoshida, S., Sasaki, N. & Suami, T. (1987) *Carbohydr. Chem.* **6**, 471.
99. Osberger, T. F. (1986) In: O'Brien Nabors, L. & Gelardi, R. C. (eds), *Alternative Sweeteners.* Marcel Dekker Inc., New York, p. 245.
100. Würsch, P. (1987) *Ernährung/Nutrition* **11**, 334.
101. Bohannon, N. V., Karam, J. H. & Forsham, P. H. (1980) *J. Am. Diet. Assoc.* **76**, 555.
102. Rumessen, J. H. & Gudmand-Hoyer, E. (1986) *Gut* **27**, 1161.
103. Hyvonen, L., Kurkela, R., Koivistoinen, P. & Raitilainen, A. (1978) *J. Food Sci.* **43**, 251.
104. U.S. 3050444 (1962) [*Chem. Abstr.* **57**, 1962, 17220].
105. Barta, J. (1986) *Elelmez. Ip.* **40**, 465.
106. Słominska, L. & Maczynski, M. (1983) *Przem. Spozyw.* **37**, 366.
107. Bujake, J. E. in rcf. 99, p. 277.
108. Young, L. S. & Long, J. E. (1982) *Chemistry of Foods and Beverages. Recent Developments.* Academic Press, New York.
109. Schwengers, D. (1976) *Z. Zuckerind. Zuckerübenbau* **29**, 614.
110. Kroyer, K. (1966) *Stärke* **18**, 311.
111. Tegge, G. & Seiler, K. (1972) *Stärke* **24**, 28.
112. Tegge, G. (1976) *Z. Lebensm.-Technol.* **27**, 265, 307.
113. U.S. 3649134 (1972) [*Chem. Abstr.* **76**, 1972, 156241].
114. U.S. 3788945 (1974) [*Chem. Abstr.* **80**, 1974, 115481].
115. Morris, C. (1984) *Food Eng.* Dec., 78.
116. Dwivedi, B. S. in ref. 99, p. 165.
117. Sicard, P. J. (1982) In: Birch, G. G. & Parker, K. G. (eds), *Nutritive Sweeteners.* Appl. Sci. Publ., London, p. 145
118. Bertrand, G. & Nitzberg, G. (1928) *Compt. Rend.* **186**, 1773.
119. Khouvine, Y. & Nitzberg, G. (1933) *Compt. Rend.* **196**, 218.
120. Isbell, H. S. & Karabinos, J. W. (1952) *J. Res. Nat. Bur. Stand.* **48**, 438.
121. Neuberg, C. & Marx, F. (1907) *Biochem. Z.* **3**, 539.
122. Humoller, F. L., Wolfrom, M. L., Lew, B. W. & Goepp, R. M. (1945) *J. Am. Chem. Soc.* **67**, 1226.
123. Ballou, G. E. (1957) *J. Am. Chem. Soc.* **79**, 165.
124. Roseman, S. (1952) *J. Am. Chem. Soc.* **74**, 4467.
125. Abdel-Akker, M. & Smith, F. (1950) *Nature* **166**, 1037.
126. Ness, R. S., Fletcher, H. G. & Hudson, C. S. (1951) *J. Am. Chem. Soc.* **73**, 4759.
127. Wright, L. W. (1974) *Chem. Technol.* **4**, 42.
128. Verksler, M. A., Petrov, Yu. I., Geling, N. G. & Klabunovskii, E. I. (1974) *Izv. Akad. Nauk USSR, Ser. Khim.* 53 (in Russian).
129. Dwivedi, B. K. (1977) In: Hood, L. R., Wardrip, E. K. & Bollenback, G. N. (eds) *Carbohydrates and Health.* AVI Publishing Co., Westport, Conn., p. 27.
130. Dwivedi, B. K. (1978) In: Dwivedi, B. K. (ed.) *Low Calorie and Special Dietary Foods.* CRC Press, West Palm Beach, Fl, p. 1.
131. Brunzell, J. D. (1978) *Diabetes Care* **1**, 223.
132. Iber, F. L. & Nasrallah, S. M. (1969) *Am. J. Med. Sci.* **258**, 80.

133. Adcock, L. A. & Gray, C. H. (1957) *Biochem. J.* **65**, 554.
134. Akgun, K. H. & Ertel, N. H. (1980) *Diabetes Care* **3**, 582.
135. Forster, H. & Mehnert, H. (1979) *Akt. Ernährung* **5**, 245.
136. Frostell, G., Keyes, P. H. & Larson, R. H. (1967) *J. Nutr.* **93**, 1.
137. Hyvoenen, L., Kurkela, R., Koivistoinen, P. & Merimaa, O. (1977) *Lebensm.-Wiss. Technol.* **10**, 316.
138. Maekinen, K. K. & Soederling, E. (1980) *J. Food Sci.* **45**, 367.
139. Touster, O. (1974) In: Sipple, H. L. & McNutt, K. W. (eds) *Sugar in Nutrition*. Academic Press, New York, p. 229.
140. Fischer, E. & Stahel, R. (1891) *Ber. Dtsch. Chem. Ges.* **24**, 528.
141. Aminoff, C., Vanninen, E. & Doty, T. E. (1978) In: Counsell, J. N. (ed.) *Xylitol*. Appl. Sci. Publ. Ltd., London, p. 1.
142. Kiss, J., D'Souza, R. & Taschner, P. (1975) *Helv. Chim. Acta* **58**, 311.
143. Holland, D. & Stoddard, J. F. (1982) *Tetrahedron Lett.* **23**, 5367.
144. Onishi, H. & Suzuki, T. (1969) *Appl. Microbiol.* **15**, 1031.
145. Demetrakopoulos, G. E. & Amos, H. (1978) *World Rev. Nutr. Diet.* **32**, 96.
146. Fratzke, A. R. & Reilly, P. J. (1977) *Proc. Biochem.* **12** (7), 27.
147. Baessler, K. H. (1978) *Pharmacol. Ther. Dent.* **3**, 85.
148. Cummings, J. H. (1981) *Gut* **22**, 763.
149. Demigne, C., Remesy, C. & Rayssiguier, Y. (1980) *Reprod. Nutr. Devel.* **20**, 1351.
150. Remesy, C. & Demigne, C. (1983) *Ann. Nutr. Metab.* **27**, 57.
151. Hayes, M. L. & Roberts, K. R. (1978) *Arch. Oral Biol.* **23**, 445.
152. Platt, D. & Werrin, S. R. (1979) *J. Dent. Res.* **58**, 1733.
153. Edwardsson, S., Birkhed, D. & Mejare, B. (1977) *Acta Odont. Scand.* **35**, 257.
154. Gallagher, J. H. C. & Fussel, S. J. (1979) *Arch. Oral Biol.* **24**, 673.
155. Imfeld, T. (1977) *Helv. Chim. Acta* **21**, 1.
156. Birkhed, D. (1978) *Caries Res.* **12**, 128.
157. Pearce, E. I. F. & Gallagher, J. H. C. (1979) *NZ Dent. J.* **75**, 8.
158. Kleber, C. J., Schimmele, R. G., Putt, M. S. & Muhler, C. (1979) *J. Dent. Res.* **58**, 614.
159. Maiwald, H.-J., Banoczy, J., Tietze, W., Toth, Z. & Wegh, A. (1982) *Zahn-, Mund-Kieferheilk.* **70**, 598.
160. WHO/FAO 22nd Report of the Joint Expert Committee on Food Additives, Geneva. *WHO Technical Report Series No. 631*, 1978, p. 29.
161. Emodi, A. (1978) *Food Technol.* **32**, 28.
162. Voirol, P. A. (1979) In: Birch, G. G. & Parker, K. J. (eds) *Sugar: Science and Technology*. Appl. Sci. Publ. Ltd., London, p. 324.
163. Voirol, P. A. (1980) In: Koivistoinen, P. & Hyvoenen, L. (eds) *Carbohydrate Sweeteners in Foods and Nutrition*. Academic Press, London, p. 269.
164. Varo, P., Westermarck, I., Rosendahl, C., Hyvoenen, L. & Koivistoinen, P. (1979) *Lebensm.-Wiss. Technol.* **12**, 153.
165. Montgomery, E. M. & Hudson, C. S. (1930) *J. Am. Chem. Soc.* **52**, 2101.
166. Gatzsche, L. & Haenel, H. (1967) *Ernährungsforschung* **12**, 641.
167. U.S. 3816174 (1974) [*Chem. Abstr.* **80**, 1974, 84755].
168. Span. 397810 (1974) [*Chem. Abstr.* **81**, 1974, 116641].
169. Demaimay, H. & Baron, C. (1978) *Lait* **234**, 575.
170. U.S. 3514327 (1970) [*Chem. Abstr.* **73**, 1970, 45791].
171. Ipatiew, W. (1912) *Ber. Dtsch. Chem. Ges.* **45**, 3218.
172. Senderens, J. B. (1920) *Compt. Rend.* **170**, 47.
173. Wolfrom, M. L., Hann, R. M. & Hudson, C. S. (1952) *J. Am. Chem. Soc.* **74**, 1105.

174. van Velthuysen, J. A. (1979) *J. Agric. Food Chem.* **27**, 680.
175. Booy, C. J. (1987) *Bull. Int. Dairy Fed.* **212**, 62.
176. Strater, P. J. in ref. 99, p. 217.
177. Schiweck, H. (1980) *Alimenta* **19**, 5.
178. Lindner, H. J. & Lichtenthaler, F. W. (1981) *Carbohydr. Res.* **93**, 135.
179. Lichtenthaler, F. W. & Lindner, H. J. (1981) *Liebigs Ann. Chem.* 2374.
180. Grupp, U. & Siebert, G. (1978) *Res. Exp. Med.* (Berl.) **173**, 261.
181. Ziesenitz, S. C. (1986) *Z. Ernährungswiss.* **25**, 253.
182. Karle, E. J. & Gehring, F. (1979) *Dtsch. Zahnärztl. Z.* **34**, 551.
183. Karle, E. J. & Gehring, F. (1981) *Dtsch. Zahnärztl. Z.* **36**, 673.
184. Gehring, F. & Karle, E. J. (1981) *Z. Ernährungwiss.* **20**, 96.
185. Siebert, G. & Ziesenitz, S. C. (1985) In: Wenk, C., Kronauer, M., Schutz, Y. & Bickel, H. (eds) *Verwertung der Nahrungsenergie durch Mensch und Tier*. Wissenschaftliche Verlagsgesellschaft, Stuttgart, p. 135.
186. Ziesenitz, S. C. (1986) In: Dölp, R. & Löhlein, D. (eds) *Aktuelle Entwicklung und Standard der künstlichen Ernährung*. Infusion, Therapy and Clinical Nutrition, Vol. 16. Karger, Basel, p. 120.
187. Siebert, G., Grupp, U. & Heinkel, K. (1975) *Nutr. Metabol.* **18** (Suppl. 1), 191.
188. Kirchgessner, M., Zinncr, P. M. & Roth, H.-P. (1983) *Int. J. Vit. Nutr. Res.* **53**, 86.
189. Taufel, A. & Taufel, K. (1970) *Nährung* **14**, 331.
190. Drost, H., Spengler, M., Gierlich, P. & Jahnke, K. (1980) *Aktuelle Endokrinologie*, Vol. 2. Georg Thieme Verlag, Stuttgart, p. 171.
191. Ziesenitz, S. C. & Siebert, G. (1987) In: Grenby, T. H. (ed.) *Developments in Sweeteners*, Vol. 3. Appl. Sci. Publ. Ltd., London, p. 109.
192. U.S. 4262032 (1981) [*Chem. Abstr.* **95**, 1981, 78771].
193. Levin, G. V. in ref. 99, p. 155.
194. Fischer, E. & Piloty, O. (1891) *Ber. Dtsch. Chem. Ges.* **24**, 521.
195. Meyer, A. S. & Reichstein, T. (1946) *Helv. Chim. Acta* **29**, 152.
196. Hudson, C. S., Hartley, O. & Purves, C. B. (1934) *J. Am. Chem. Soc.* **56**, 1248.
197. Austin, W. S. & Humoller, F. L. (1934) *J. Am. Chem. Soc.* **56**, 1153.
198. Steiger, M. & Reichstein, T. (1935) *Helv. Chim. Acta* **18**, 790.
199. Glatthaar, C. & Reichstein, T. (1938) *Helv. Chim. Acta* **21**, 3.
200. Mottern, H. H. & Cole, H. L. (1939) *J. Am. Chem. Soc.* **61**, 2701.
201. Pigman, W. W. (1940) *J. Res. Natl. Bur. Stand.* **25**, 301.
202. Isbell, H. S. & Frush, H. L. (1944) *J. Res. Natl. Bur. Stand.* **32**, 77.
203. Wolfrom, M. L. & Thompson, A. (1946) *J. Am. Chem. Soc.* **68**, 791.
204. Grenby, T. H. (1983) In: Grenby, T. H., Parker, K. J. & Lindley, M. G. (eds), *Developments in Sweeteners*, Vol. 2. Appl. Sci. Publ. Ltd., London, p. 51.
205. Siebert, G., Romen, W., Schnell-Dompert, E. & Hannover, R. (1980) *Infusionsther. Klin. Ernähr.* **7**, 271.
206. Hers, H. G. (1952) *Biochim. Biophys. Acta* **8**, 416.
207. Leuthardt, F. & Testa, E. (1950) *Helv. Chim. Acta* **33**, 1919.
208. Sanchez, J. J., Gonzales, N. S. & Pontis, H. G. (1971) *Biochim. Biophys. Acta* **227**, 67.
209. Lardy, H. A., Wiebelhaus, V. D. & Mann, K. M. (1950) *J. Biol. Chem.* **187**, 325.
210. Würsch, P., Welsch, C. & Arnavid, M. J. (1979) *Nutr. Metab.* **24**, 145.
211. Cauley, L. (1987) *The Washington Times* Nov. 25, 5.
212. Czech. 224984 (1984) [*Chem. Abstr.* **102**, 1985, 222434]; Czech. 225526 (1984) [*Chem. Abstr.* **103**, 1985, 37702]; Czech. 231544 (1986) [*Chem. Abstr.* **106**, 1987, 196729].

213. Daniell, W. F. (1855) *Pharmacol. J.* **14**, 158.
214. Inglett, G. E. & May, J. F. (1968) *Econ. Bot.* **22**, 326.
215. van der Wel, H. & Loeve, K. (1972) *Eur. J. Biochem.* **31**, 221.
216. Higginbotham, J. D. (1979) In: Birch, G. G., Brennan, J. G. & Paker, K. J. (eds) *Sensory Properties of Foods*. Appl. Sci. Publ. Ltd., London, p. 87.
217. Higginbotham, J. D. (1977) In: Birch, G. G., Brennan, J. G. & Parker, K. J. (eds) *Sensory Properties of Foods*. Appl. Sci. Publ. Ltd., London, p. 129.
218. Iyengar, R. B., Smits, P., van der Ouderaa, F. J. G., van der Wel, H., van Brouwershaven, J., Ravenstein, P., Richters, G. & van Wassenaar, P. D. (1979) *Eur. J. Biochem.* **96**, 193.
219. van der Wel, H., Iyengar, R. B., van Brouwershaven, J., van Wassenaar, P. D., Bel, W. J. & van der Ouderaa, F. J. G. (1984) *Eur. J. Biochem.* **144**, 41.
220. van der Wel, H. & Bel, W. J. (1980) *Eur. J. Biochem.* **104**, 413.
221. Edens, L. & van der Wel, H. (1985) *Trends Biotechnol.* **3**, 61.
222. Kim, S. H., de Vos, A. & Ogata, C. (1988) *Trends Biochem. Sci.* **13**, 13.
223. Korver, O., van Gorkom, M. & van der Wel, H. (1973) *Eur. J. Biochem.* **35**, 554.
224. van der Wel, H. (1986) In: Hudson, B. J. P. (ed.) *Developments in Food Proteins*, Vol. 4. Appl. Sci. Publ. Ltd., London, p. 219–245.
225. O'Brien Nabors, L. & Gelardi, R. C. (eds) (1986) *Alternative Sweeteners*. Marcel Dekker, New York, Basel.
226. Kinghorn, A. D. & Soejarto, D. D. (1986) *CRC Critical Reviews in Plant Sciences*, Vol. 4. CRC Press, Boca Raton, Fl., p. 79.
227. Kinghorn, A. D. & Soejarto, D. D. (1989) *Med. Res. Revs.* **9**, 91.
228. Irvine, F. R. (1961) *Woody Plants of Ghana*. Oxford University Press, London.
229. Inglett, G. E. & May, J. F. (1969) *J. Food Sci.* **34**, 408.
230. Morris, J. A. & Cagan, R. H. (1972) *Biochim. Biophys. Acta* **261**, 114.
231. van der Wel, H. (1972) *FEBS Lett.* **21**, 88.
232. Morris, J. A., Martenson, R., Deibler, G. J. & Cagan, R. H. (1973) *J. Biol. Chem.* **248**, 534.
233. van der Wel, H. & Loeve, K. (1973) *FEBS Lett.* **29**, 181.
234. Frank, G. & Zuber, H. (1976) *Hoppe Seyler's Z. Physiol. Chem.* **357**, 586.
235. Bohak, Z. & Li, S. L. (1976) *Biochim. Biophys. Acta* **427**, 153.
236. Frank, G. & Zuber, H. (1976) *Hoppe Seyler's Z. Physiol. Chem.* **357**, 585.
237. Brand, J. G. & Cagan, R. H. (1977) *Biochim. Biophys. Acta* **493**, 178.
238. Brand, J. G., Cagan, R. H. & Bayley, D. L. (1985) *Proc. Soc. Exp. Biol. Med.* **179**, 76.
239. van der Wel, H. (1980) *Trends Biochem. Sci.* **5**, 122.
240. Hellekant, G. (1976) *Chem. Senses Flavor* **2**, 97.
241. Brit. 2123672 (1984) [*Chem. Abstr.* **101**, 1984, 22240].
242. Furukawa, S. & Tomizawa, Z. (1920) *J. Chem. Ind. Tokyo* **23**, 342 [*Chem. Abstr.* **14**, 1920, 2839].
243. Tsuzuki, Y. (1970) *J. Chem. Educ.* **47**, 695.
244. Acton, E. M., Stone, H., Leaffer, M. A. & Oliver, S. M. (1970) *Experientia* **25**, 173.
245. Acton, E. M. & Stone, H. (1976) *Science* **193**, 584.
246. Settzer, R. J. (1975) *Chem. Eng. News* 27.
247. Crammer, B. & Ikan, R. (1977) *Chem. Soc. Rev.* **5**, 431.
248. Mitoma, C., Acton, E. M., DeGraw, J. I. & Thomas, D. M. (1980) *Drug Chem. Toxicol.* **8**, 195.
249. Iwamura, H. (1980) *J. Med. Chem.* **23**, 308.
250. Venanzi, T. J. & Venanzi, C. A. (1988) *J. Med. Chem.* **31**, 1879.

251. Compadre, C. M., Pezzutto, J. M., Kinghorn, A. D. & Kamath, S. K. (1985) *Science* **227**, 417.
252. Compadre, C. M., Hussain, R. A., Lopez de Compadre, R. L., Pezzutto, J. M. & Kinghorn, A. D. (1987) *J. Agric. Food Chem.* **35**, 273.
253. Mori, K. & Kato, M. (1986) *Tetrahedron* **42**, 5895.
254. Fullas, F., Kim, J., Compadre, C. M. & Kinghorn, A. D. (1989) *J. Chromatogr.* **464**, 213.
255. Wood, J. H. B., Allerton, R., Diehl, H. W. & Fletcher, H. G. J. (1955) *J. Org. Chem.* **20**, 875.
256. Kohda, H., Kasai, R., Yamasaki, K., Murakami, K. & Tanaka, O. (1976) *Phytochemistry* **15**, 981.
257. Sakamoto, I., Yamasaki, K. & Tanaka, O. (1977) *Chem. Pharm. Bull.* **25**, 844.
258. Makapugay, H. C., Nanayakkara, N. P. D. & Kinghorn, A. D. (1984) *J. Chromatogr.* **283**, 390.
259. Darise, M., Kohda, H., Mizutani, K., Kasai, R. & Tanaka, O. (1983) *Agric. Biol. Chem.* **47**, 133.
260. Japan 63177764 (1987) [*Chem. Abstr.* **110**, 1989, 230444].
261. Yamasaki, K., Kohda, H., Kobayashi, T., Kasai, R. & Tanaka, O. (1976) *Tetrahedron Lett.* 1005.
262. Mosettig, E. & Nes, W. R. (1955) *J. Org. Chem.* **20**, 891.
263. Mosettig, E., Berlinger, U., Dolder, F., Lichti, H., Quitt, P. & Waters, J. A. (1963) *J. Am. Chem. Soc.* **85**, 2305.
264. Japan 8019009 [*Chem. Abstr.* **93**, 1980, 4133].
265. Hsing, Y. I., Su, W. F. & Chang, W. C. (1983) *Bot. Bull. Acad. Sin.* **24**, 115 [*Chem. Abstr.* **99**, 1983, 17877].
266. Ogawa, T., Nozaki, M. & Matsui, M. (1980) *Tetrahedron* **36**, 2641.
267. Ruddat, M., Heffmann, E. & Lang, A. (1965) *Arch. Biochem. Biophys.* **110**, 496.
268. Kinghorn, A. D., Nanayakkara, N. P. D., Soejarto, D. D., Medon, P. J. & Kamath, S. (1982) *J. Chromatogr.* **237**, 478.
269. DuBois, G. E., Dietrich, P. S., Lee, J. F., McGarraugh, G. V. & Stepenson, R. A. (1981) *J. Med. Chem.* **24**, 1269.
270. van der Wel, H., van der Heijden, A. & Peer, H. G. (1987) *Food Rev. Int.* **3**, 193.
271. DuBois, G. E. & Stephenson, R. A.(1985) *J. Med. Chem.* **28**, 93.
272. Esaki, S., Tanaka, R. & Kamiya, S. (1984) *Agric. Biol. Chem.* **48**, 1831.
273. Tanaka, O. (1982) *Trends Anal. Chem.* **1**, 246.
274. Pezzutto, J. M., Compadre, C. M., Swanson, S. M., Nanayakkara, N. P. D. & Kinghorn, A. G. (1985) *Proc. Natl. Acad. Sci.* **82**, 2478.
275. Pezzutto, J. M. (1986) In: Atta-ur-Rahman & Le Quesne, P. W. (eds) *New Trends in Natural Products Chemistry 1986*. Elsevier Science Publishers, Amsterdam, p. 371.
276. Kinghorn, A. D. & Compadre, C. M. (1985) *Pharm. Int.* **6**, 201.
277. Tanaka, T., Kohda, H., Tanaka, O., Chen, F. H., Chou, W. H. & Leu, J. L. (1981) *Agric. Biol. Chem.* **45**, 2165.
278. Darise, M., Mizutani, K., Kasai, R., Tanaka, O., Kitahata, S., Okada, S., Ogawa, S., Murakami, F., Chen, F. H. (1984) *Agric. Biol. Chem.* **48**, 2483.
279. Tanaka, T., Tanaka, O., Lin, Z. W., Zhou, J. & Ageta, H. (1983) *Chem. Pharm. Bull.* **31**, 780.
280. Tanaka, T., Tanaka, O., Lin, Z. W. & Zhou, J. (1985) *Chem. Pharm. Bull.* **33**, 4275.
281. Tahara, A., Nakata, T. & Ohtsuka, Y. (1971) *Nature* **233**, 619.
282. Tschirch, A. & Cederberg, H. (1907) *Arch. Pharm.* **245**, 97.

283. Voss, W. & Pfirschke, J. (1937) *Chem. Ber.* **70**, 132.
284. Beaton, J. M. & Spring, F. S. (1955) *J. Chem. Soc.* 3126.
285. Hodge, J. E. & Inglett, G. E. (1974) In: Inglett, G. E. (ed.) *Symposium: Sweeteners.* Avi Publishing Company, Inc., Westport, Conn., Chapter 20.
286. Cook, M. K. & Gorminger, B. H. (1974) In: Inglett, G. E. (ed.) *Symposium: Sweeteners.* Avi Publishing Company, Inc., Westport, Conn., Chapter 19.
287. Sasaki, Y., Morita, T., Kuramoto, T., Mizutani, K., Ikeda, R. & Tanaka, O. (1988) *Agric. Biol. Chem.* **52**, 207.
288. Esaki, S., Konishi, F. & Kamiya, S. (1978) *Agric. Biol. Chem.* **42**, 1599.
289. Brieskorn, C. H. & Lang, J. (1978) *Arch. Pharm.* **311**, 1001.
290. Louis, L. H. & Conn, J. W. (1956) *J. Lab. Clin. Med.* **47**, 20.
291. Kitagawa, I., Sakagami, M., Hashiuchi, F., Zhou, J. L., Yoshikawa, M. & Ren, J. (1989) *Chem. Pharm. Bull.* **37**, 551.
292. Hashimoto, Y., Ishizone, H., Suganuma, M., Ogura, M., Nakatsu, K. & Yoshioka, H. (1983) *Phytochemistry* **23**, 259.
293. Hashimoto, Y., Ishizone, H. & Ogura, M. (1980) *Phytochemistry* **19**, 2411.
294. Hashimoto, Y., Ohta, Y., Ishizone, H., Kuriyama, M. & Ogura, M. (1982) *Phytochemistry* **21**, 2335.
295. Makapugay, H. C., Nanayakkara, N. P. D., Soejarto, D. D. & Kinghorn, A. D. (1985) *J. Agric. Food Chem.* **33**, 348.
296. Lee, C. H. (1975) *Experientia* **31**, 533.
297. Takemoto, T., Arihara, S., Nakajima, T. & Okuhira, M. (1983) *Yakugaku Zasshi* **103**, 1151, 1155, 1167.
298. Kasai, R., Matsumoto, K., Nie, M. R. L., Morita, T., Awazu, A., Zhou, J. & Tanaka, O. (1987) *Phytochemistry* **26**, 1371.
299. Jizba, J. & Herout, V. (1967) *Collect. Czech. Chem. Commun.* **32**, 2867.
300. Kresánek, J. & Krejča, J. (1977) *Atlas of Healing Plants and Forest Fruits.* Osveta, Martin (in Slovak).
301. Jizba, J., Dolejs, L., Herout, V. & Sorm, F. (1971) *Tetrahedron Lett.* 1329.
302. Havel, M. & Cerny, V. (1975) *Collect. Czech. Chem. Commun.* **40**, 1579.
303. Reisch, J. & Dawidar, A. M. (1978) *Sci. Pharm.* **46**, 281.
304. Kim, J., Pezzutto, J. M., Soejarto, D. D., Land, F. A. & Kinghorn, A. D. (1988) *J. Nat. Prod.* **51**, 1166.
305. Kim, J. & Kinghorn, A. D. (1989) *Phytochemistry* **28**, 1225.
306. Arakawa, H. & Nakazaki, M. (1959) *Chem. Ind.* (London) 671.
307. Asahina, Y. & Asano, J. (1929) *Chem. Ber.* **62**, 171 and ibid. **64** (1931) 1252.
308. Suzuki, H., Ikeda, T., Matsumoto, T. & Noguchi, M. (1977) *Agric. Biol. Chem.* **41**, 719.
309. Takeuchi, N., Murase, M., Ochi, K. & Tobinaga, S. (1980) *Chem. Pharm. Bull.* **28**, 3613.
310. Takeuchi, N., Murase, M., Ochi, K. & Tobinaga, S. (1975) *J. Chem. Soc. Chem. Commun.* 820.
311. Naoi, Y., Higuchi, S., Ho, H., Nakano, T., Sakai, K., Matsui, T., Wagatsuma, S., Nishi, A. & Sano, S. (1975) *Org. Prep. Proc. Int.* **7**, 129.
312. Dick, W. E. J. (1981) *J. Agric. Food Chem.* **29**, 305.
313. Yamato, M., Hashigaki, K., Honda, E., Sato, K. & Koyama, T. (1977) *Chem. Pharm. Bull.* **25**, 695.
314. Yamato, M., Hashigaki, K., Mito, K. & Koyama, T. (1978) *Chem. Pharm. Bull.* **26**, 2321.
315. Dick, W. W. J. & Hodge, J. E. (1978) *J. Agric. Food Chem.* **26**, 723.

316. Arnoldi, A., Camarda, L. & Merlini, L. (1986) *J. Agric. Food Chem.* **34**, 339.
317. Horowitz, R. M. (1986) In: Cody, V., Middleton, E., Jr. & Harborne, J. B. (eds) *Plant Flavonoids in Biology and Medicine: Biochemical, Pharmacological and Structure–Activity Relationships*. Alan R. Liss, New York, p. 163.
318. Konishi, F., Esaki, S. & Kamiya, S. (1983) *Agric. Biol. Chem.* **47**, 265.
319. Horowitz, R. M. & Gentili, B. (1974) In: Inglett, G. E. (ed.), *Symposium: Sweeteners*. Avi Publishing Company, Inc., Westport, Conn., Chapter 16.
320. Guadagni, D., Maier, V. P. & Turnbaugh, J. C. (1973) *J. Sci. Food Agric.* **24**, 1277.
321. Gaffield, W., Lundin, R. E., Gentili, B. & Horowitz, R. M. (1975) *Bioorg. Chem.* **4**, 259.
322. Hoerhammer, L., Wagner, H., Roestler, M., Keckeisen, M. & Farkas, L. (1965) *Tetrahedron* **21**, 969.
323. U.S. 3087821 (1963) [*Chem. Abstr.* **59**, 1963, 1780].
324. Horowitz, R. M. & Gentili, B. (1969) *J. Agric. Food Chem.* **17**, 696.
325. Horowitz, R. M. & Gentili, B. (1961) *Arch. Biochem. Biophys.* **92**, 191.
326. Horowitz, R. M. & Gentili, B. (1963) *Tetrahedron* **19**, 773.
327. U.S. 3583894 (1971) [*Chem. Abstr.* **75**, 1971, 47652].
328. U.S. 3375242 (1968) [*Chem. Abstr.* **69**, 1968, 52456].
329. Krbechek, L., Inglett, G., Holik, M., Dowling, B., Wagner, R. & Riter, R. (1968) *J. Agric. Food Chem.* **16**, 108.
330. DuBois, G. E., Crosby, G. A. & Saffron, P. (1977) *Synth. Commun.* **7**, 49.
331. DuBois, G. E., Crosby, G. A. & Stephenson, R. A. (1981) *J. Med. Chem.* **24**, 408.
332. Arakawa, H. & Nakazaki, M. (1960) *Ann. Chem.* **636**, 111.
333. Wingard, R. E. J., Crosby, G. A. & DuBois, G. E. (1978) *CHEMTECH* Oct., 616
334. DuBois, G. E., Crosby, G. A. & Saffron, P. (1977) *Science* **195**, 397.
335. DuBois, G. E. & Stephenson, R. A. (1982) *J. Agric. Food Chem.* **30**, 476.
336. Crammer, B. & Ikan, R. (1977) *Chem. Soc. Revs.* **6**, 431.
337. Beerens, H. (1981) *Ann. Fals. Exp. Chim.* **74**, 261.
338. Hung. 4026 (1970) [*Chem. Abstr.* **77**, 1981, 60321].
339. U.S. 3826856 (1974) [*Chem. Abstr.* **82**, 1975, 299923].
340. U.S. 3876777 (1975) [*Chem. Abstr.* **83**, 1975, 28539].
341. Konishi, F., Esaki, S. & Kamiya, S. (1983) *Agric. Biol. Chem.* **47**, 1419.
342. Chinn, L. J., Wakters, D. E., Williams, K. & Mazur, R. H. (1987) *J. Agric. Food Chem.* **35**, 409.
343. Wong, R. Y. & Horowitz, R. M. (1986) *J. Chem. Soc., Perkin Trans. I* 843.
344. U.S. 4031260 (1977) [*Chem. Abstr.* **87**, 1977, 83375].
345. Gentili, B. & Horowitz, R. M. (1971) *J. Chromatogr.* **63**, 467.
346. Nanayakkara, N. P. D., Hussain, R. A., Pezzutto, J. M., Soejarto, D. D. & Kinghorn, A. D. (1988) *J. Med. Chem.* **31**, 1250.
347. Kasai, R., Hirono, S., Chou, W. H., Tanaka, O. & Chen, F. H. (1988) *Chem. Pharm. Bull.* **36**, 4167.
348. Daniel, W. F. (1852) *Pharmacol. J.* **11**, 445.
349. Inglett, G. E., Dowling, B., Albrecht, J. J. & Hoglan, F. A. (1965) *J. Agric. Food Chem.* **13**, 284.
350. Kurihara, K. & Beidler, L. M. (1968) *Science* **161**, 1241.
351. Brouwer, J. N., van der Wel, H., Francke, A. & Hanning, G. J. (1968) *Nature* **220**, 373.
352. Kurihara, Y. & Terasaki, S. (1982) *Biochim. Biophys. Acta* **719**, 444.
353. Giroux, E. L. & Henkin, R. I. (1974) *J. Agric. Food Chem.* **22**, 595.
354. Cagan, R. H. (1973) *Science* **181**, 32.

355. Kurihara, K. & Beidler, L. M. (1969) *Nature* **222**, 1176.

356. Dzendolet, E. (1969) *Percept. Psychophys.* **6**, 187.

357. Bartoshuk, L. M., Gentile, R. L., Moskowitz, H. R. & Meiselman, H. L. (1974) *Physiol. Behavior* **12**, 449.

358. Brouwer, J. N., Glaser, D., Hard Af Segerstad, C., Hellekant, G., Ninomiya, Y. & van der Wel, H. (1983) *J. Physiol.* (Lond.) **337**, 221.

359. Bartoshuk, L. M., Lee, C. H. & Scarpellino, R. (1972) *Science* **178**, 988.

360. Panizzi, L. & Scarpati, M. L. (1954) *Nature* **174**, 1062.

361. Warren, R. P., Warren, R. M. & Weniger, M. G. (1969) *Nature* **233**, 94.

362. Hooper, D. (1887) *Pharm. J. Trans.* **17**, 867.

363. Stöklin, W. E. (1969) *Helv. Chim. Acta* **50**, 491.

364. Kennedy, L. M., Saul, L. R., Sefecka, R. & Stevens, D. A. (1988) *Chem. Senses* **13**, 529.

365. Kolodny, D. E. & Kennedy, L. M. (1988) *Chem. Senses* **13**, 545.

3

Synthetic sweeteners

Synthetic sweeteners were discovered accidentally in June 1878 when Fahlberg, after picking up a weighing paper, licked his finger and found it tasted sweet. The compound responsible for the sweet taste was saccharin — the oldest known, and still used artificial sweetener. In 1884 another synthetic sweetener was identified and characterized — 4-ethoxyphenylurea, and since then hundreds of compounds with sweet taste have been synthesized.

The search for the evaluation of physiological and toxicological properties of synthetic sweeteners was carried out in parallel on all new discoveries by synthetic organic chemists. Only saccharin was not tested in this manner: instead, Fahlberg himself tested it for toxicity by consuming 10 grams of saccharin before recommending it for commercial exploitation.

It was not until the second half of the 20th century that another artificial sweetener — cyclamate — was discovered, once again by chance in 1937, and introduced into commercial use. When used either alone or together with saccharin, it produced 'better' or 'more natural' taste. Serendipity was involved in the discovery of other synthetic sweeteners. Thus in 1966 Schlatter discovered a sweet dipeptide — aspartame and, in 1967 Clauss came across the sweet taste of oxathiazinone dioxides.

The long history of saccharin illustrates the changing status of the novel sweetener. At the time of its introduction on the market it was considered a cheap substitute for sucrose extracted from sugar beet or sugar cane. Its consumption and its status rose sharply during both World Wars as the natural precursor for sucrose manufacture was in short supply.

Its status has been boosted latterly by the growing number of findings that the consumption of natural sucrose is not necessarily beneficial to health. An excessive intake of sucrose was identified as being associated with obesity and increased blood pressure. Furthermore, it stimulated the progress of arteriosclerosis, a prime cause of heart ischaemic disease, as well as coronary brain disorders. Thus artificial sweeteners became an integral part of our efforts to minimize the energy uptake in human organism by proper diet.

The role of artificial sweeteners as a substitute for sucrose continues to be a topic of debate. The revival of interest in research into synthetic sweeteners was prompted by economic factors (sucrose is a cheap food additive in many countries, but its price on world markets has lost some of its stability as sucrose became a starting material for both commercial bio(techno)logical and chemical processes) and because of health considerations. A reduction in human sugar consumption is recommended for many reasons, the prevention of dental caries and obesity being probably the most compelling.

Although both sucrose and artificial sweeteners are a part of the human diet, their status and role differ markedly. Whilst sugars are considered as a constituent of food, synthetic sweeteners are low-calorie sensory bioactive compounds* used mainly in diet foods. They are also used in pharmaceutical applications to mask the unpleasant taste of the drug or as a substitute for sugar (sucrose) for diabetes patients [2], as well as in cosmetic preparations for mouth hygiene.

Many strict requirements must be satisfied before an artificial sweetener can be accepted, the most important is that it must not pose a health hazard. It must be free from any toxic effects and should not interfere with the normal functions of the human organism, even in the long-term use. The use and application of artificial sweeteners is regulated by the international committees of WHO and FAO for food additives which legislate their maximal daily dose [4–8]. The final provisions for the commercial use of novel synthetic sweeteners, such as Aspartame and Acesulfam K, have been agreed; only acceptable daily intake (ADI) values have been recommended for saccharin and cyclamate [3], further tests and investigations are being made before any unequivocal solution is found.

---

* Artificial sweeteners are defined as compounds with substantially higher sweetness than sucrose, but with zero or negligible energy content [1].

A vital criterion of all synthetic sweeteners is the measure of the intensity and quality (clean and pleasant) of the elicited sweet taste. In order to be successful in food applications, the response must be elicited in 1 to 2 seconds. In general, the intensity of the sweet taste is proportional to the concentration and the quality is measured by a comparison with a dilute solution of sucrose. When subjected to the above comparisons, aspartame appears to be the best and saccharin the worst. However, the most acceptable results are usually achieved by a combination of two or more sweeteners.

Apart from the subjective sensory responses sweet substances must also possess suitable physico-chemical properties, and their production has to be simple and cheap.

Furthermore, they must be chemically inert towards all natural components as well as towards all other artificial additives; they should have sufficient thermal and photochemical stability for their intended application. This is especially truc if the sweeteners are expected to survive many technological processes in the food industry, such as heating, high pressure, freezing and lyophilization — their thermal stability range should be from −30 to 260 °C.

The sweeteners must be also stable in solution over a wide range of acidity, starting from pH 2.5 (acidic soft drinks) to pH 8.

Thus it should not be surprising that from the vast numbers of so far synthesized sweet tasting compounds, only a few have passed all the required tests. Even a commercially successful sweetener, such as saccharin, is limited in its use due to its faint bitter taste. Several sweeteners, used over a period of time, were later found to be toxic and their use prohibited (dulcin, P-4000, perillaldoxime).

To date, there is no ideal sweetener which can match sucrose in its sensory and functional properties, have no harmful effects on the tooth enamel, be suitable for diabetics, have low-calorie value and be easily and cheaply produced. Real sweeteners are characterized by various combinations of the above deficiencies (from the point of view of possible health hazard, sensory properties, technological problems in their production, costly production and other). In other words, each synthetic sweetener is disqualified from certain applications, but a single or mixed sweetener exists for each specific application. Any future potential synthetic sweetener will be scrutinized against all the above-mentioned conditions — a very costly exercise in order to prove that its use does not have any health risks.

Table 3.1 summarizes data on the use of saccharin, cyclamate, aspartame and Acesulfam K in some European countries. It is evident that at least three different synthetic sweeteners are allowed in most

Table 3.1 — Utilization of some sweeteners in Europe

Country	Saccharin			Cyclamate			Aspartame			Acesulfam K		
	P	N	S	P	N	S	P	N	S	P	N	S
Belgium	a	a	a	c	c	a	c	a	a	c	a	a
CSFR	a	a	a	a	a	a		a	a			
Denmark	a	a	a	c	a	a	a	a	a	c	a	a
Finland	a	a	a	a	a	a	c	a	a			
France	c	c	a	c	c	a	c	c	a			
Greece	c	c	a	c	c	a	c	c	a			
Holland	a	a	a	c	a	a	b	b	a	b	b	b
Luxembourg	a	a	a	a	a	a	a	a	a			
Norway	a	a	a	a	a	a	a	a	a			
FRG	a	a	a	a	a	a	a	b	a			a
Portugal	c	c	a	c	c	a	c	c	a			
Austria	a	a	a	c	c	a	c	c	a			
Spain	a	a	a	a	a	a	b	b	b			
Switzerland	a	a	a	a	a	a	a	a	a	a	a	a
Sweden	a	a	a	c	c	a	a	a	a	b	b	b
Italy	a	a	a	a	a	a	c	c	a	c	c	a
Great Britain	a	a	a	c	c	c	a	a	a	a	a	a

In the former Soviet Union and in Bulgaria Acesulfam K has been approved for use in tooth pastes [78].
P — food constituent, N — drink constituent, S — tabletop sweetener. a — licensed, b — awaiting a license, c — forbidden.

countries, thus giving the food industry enough scope for mixing them into a number of composite, proprietary and specific sweeteners.

3.1 SACCHARIN

Saccharin (3-oxo-2,3-dihydrobenzo[*d*]isothiazole-1,1-dioxide), the oldest known synthetic sweetener, is 300–500 times sweeter than sucrose. It was discovered accidentally by the German chemist Constantine Fahlberg in 1878 during his work on the oxidation of *o*-toluenesulphonamide in the laboratory of Professor Remsden at Johns Hopkins University. He immediately recognized the significance of his discovery and patented the synthetic method. He published it later both in Germany and the United States [9, 10]. The commercial production of saccharin started in 1884.

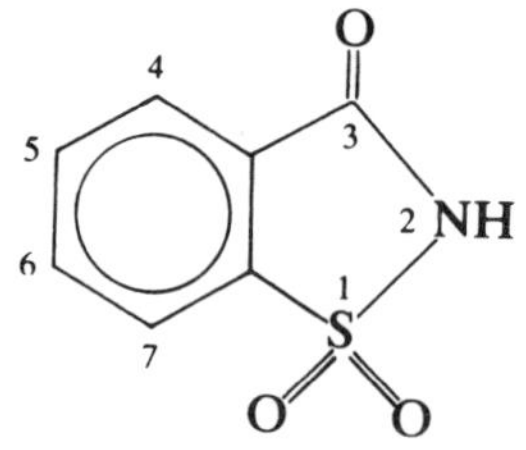

Saccharin

Saccharin is a white, crystalline compound with a melting point of 228–229 °C. One gram of saccharin dissolves in 290 ml of cold or in 25 ml of hot water, in 50 ml of glycerol, 31 ml of ethanol and in as little as 12 ml of acetone; an addition of citric, tartaric or acetic acid improves its water solubility. It is practically insoluble in chloroform, but can be successfully extracted into ether or petroleum ether. Saccharin itself is quite a strong acid ($pK_a = 1.3$) and dissolves well in solutions of alkali hydroxides and carbonates. It has a tendency to dimerize but remains stable for an hour in aqueous buffers at pH 3.3, 7.0, 9.0 and at 150 °C. Decomposition is observed only under extreme conditions. In acidic pH (under 3) saccharin hydrolyses to the ammonium salt of 2-sulphobenzoic acid; alkaline hydrolytic conditions yield the salt of 2-carboxybenzenesulphonamide.

Saccharin forms salts which are as sweet as the parent compound. The frequently used sodium salt crystallizes with two molecules of water. Drying at 120 °C for 4 hours furnishes the anhydrous salt which is valuable for its solubility in water (1 g in 1.5 ml of water at 22 °C, or in 50 ml of ethanol). This outstanding property makes it the most used commercial preparation as it is 500 times sweeter than the parent compound, saccharin. Aqueous solutions of salts are neutral or weakly alkaline and are stable in the pH range of 3.3–7.0. The calcium salt, a white crystalline and intensely sweet substance, is less sweet than the sodium salt and crystallizes with 3.5 molecules of water. The anhydrous salt, prepared analogously as the sodium salt, is very soluble in water (1 g in 1.5 ml) and 33% aqueous ethanol.

Saccharin also forms an aluminium salt [11], marginally sweet nickel salts and copper salt which causes stomach disorder [12]. X-ray crystallographic studies of saccharin revealed its almost planar structure; the crystal is composed of centrosymmetric dimeric molecules bound by N–H... O hydrogen bonds between the imidic NH group and (5-membered) ring oxygen [32, 33].

Figure 3.1 shows the Newman projection along the N–S bond of saccharin.

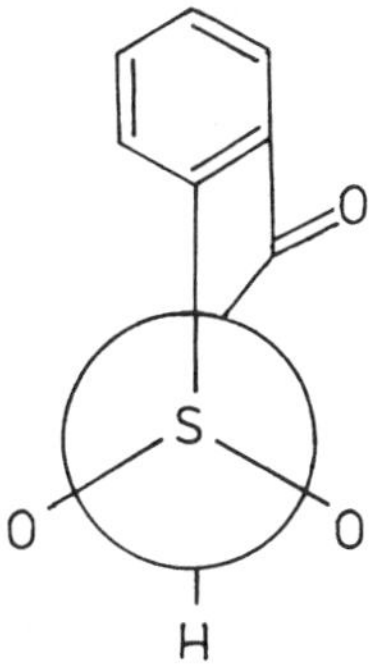

Fig. 3.1 — Newman projection along the N—S bond in saccharin, torsional angle 59°, determined by three-dimensional X-ray analysis [33].

3.1.1 Synthetic procedures leading to saccharin

The synthesis of saccharin can start from any of the four starting compounds — toluene, phthalic anhydride or phthalic acid and 2-chlorotoluene [12].

The best known method still appears to be that of Fahlberg [13]. The reaction of toluene with sulphonyl chloride yields a mixture of *o*- and *p*-toluenesulphonyl chlorides from which the *o*-isomer is separated by low-temperature centrifugation, and then converted by reaction with ammonia into *o*-methylbenzenesulphonamide. The next step involves an oxidation, usually with potassium permanganate, sodium dichromate or with solutions of cobalt or manganese salts in the presence of oxygen; acetic acid is usually used as the solvent [16]. Electrolytic oxidation can be also carried out either on its own or in conjunction with the chemical methods. The resultant carboxylic acid, the immediate precursor of saccharin, is then converted into it by cyclo-dehydration (Scheme 3.1).

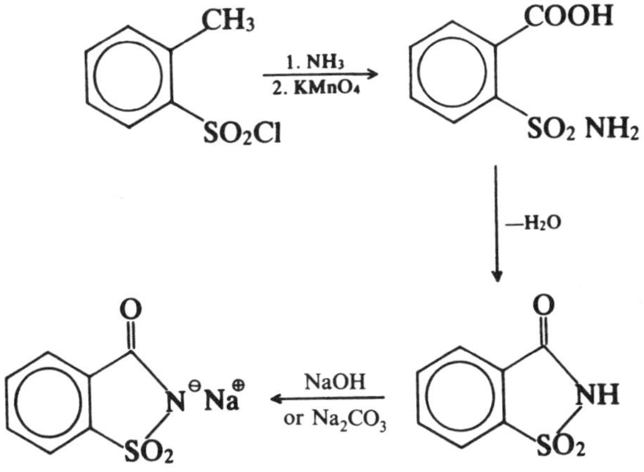

CH_3, SO_2Cl → (1. NH_3; 2. $KMnO_4$) → $COOH$, SO_2NH_2 → ($-H_2O$) → NH, SO_2, O → (NaOH or Na_2CO_3) → $N^{\ominus}Na^{\oplus}$, SO_2, O

Scheme 3.1

The by-product from the first step, *p*-toluenesulphonyl chloride, is also utilized as the starting material for the preparation of chloramine T or as the well-known tosylating reagent.

Another synthetic approach to saccharin utilizes methyl *o*-anthranilate which is prepared in three steps from phthalic anhydride [14] (Maumee process [15]).

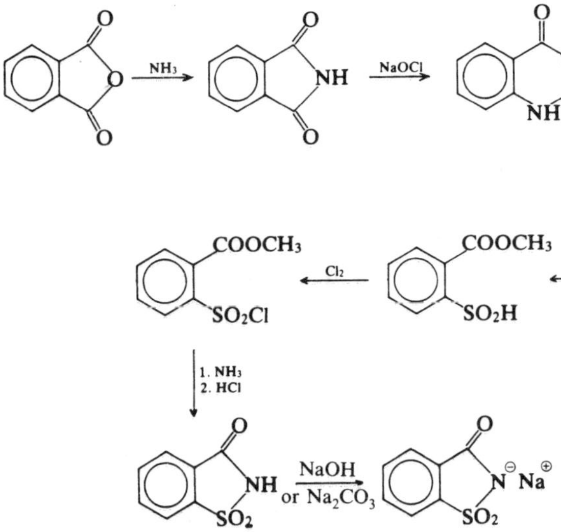

NH_3 NH NaOCl O NH O CH_3OH or CO_2 $COOCH_3$ NH_2 $NaNO_2$ HCl $COOCH_3$ SO_2Cl Cl_2 $COOCH_3$ SO_2H SO_2 $COOCH_3$ $N^{\oplus}{\equiv}N\ Cl^{\ominus}$ 1. NH_3 2. HCl NH SO_2 NaOH or Na_2CO_3 $N^{\ominus}\ Na^{\oplus}$ SO_2

Scheme 3.2

Calcium salt of saccharin was prepared by an addition of calcium carbonate into a warm suspension of saccharin (50–55 °C, pH 4.5–5.75) [17].

Derivatives of saccharin with substituents in the aromatic nucleus are accessible by the above methods as is shown in Scheme 3.3 for 6-aminosaccharin [18].

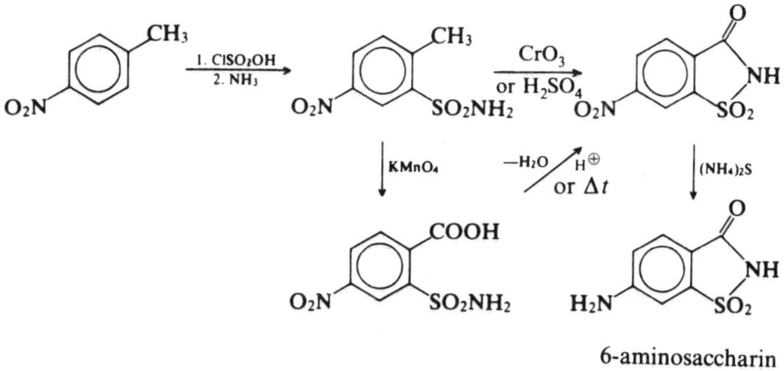

CH_3 O_2N 1. $ClSO_2OH$ 2. NH_3 CH_3 O_2N SO_2NH_2 CrO_3 or H_2SO_4 O NH O_2N SO_2 $KMnO_4$ $-H_2O$ $H^{\oplus}$ or Δt $(NH_4)_2S$ COOH O_2N SO_2NH_2 O NH H_2N SO_2

6-aminosaccharin

Scheme 3.3

The preparations of other substituted saccharins have been described [15]. The derivatives with substituents in positions 4 and 6 of the aromatic ring are, in general, sweet tasting; however, positions 5 and 7 impart bitterness. Table 3.2 lists a number of examples.

Table 3.2 — Relative sweetness of saccharin derivatives

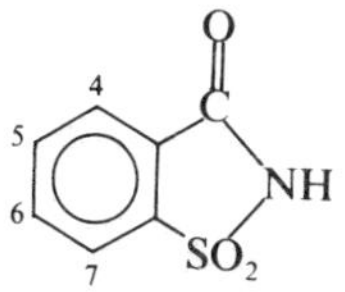

R^4	R^5	R^6	R^7	Relative sweetness	References
H	H	H	H	300–500	[12]
Cl	H	H	H	225	[19, 20]
NO_2	H	H	H	faintly sweet	[13, 19–21]
NH_2	H	H	H	sweet	[13, 20, 21]
H	Cl	H	H	faintly sweet, bitter	[19]
H	H	Cl	H	sweet	[13, 19, 20]
H	H	F	H	450	[13, 19, 20]
H	H	Br	H	sweet	[19, 20]
H	H	CH_3	H	200	[19, 20]
H	H	NO_2	H	faintly sweet	[13, 20, 21]
H	H	NH_2	H	sweet	[19, 20]
H	H	OH	H	sweet	[19, 20]
H	II	SO_2NH_2	H	sweet	[19, 21]
H	H	H	Cl	sweet, bitter	[19]

^{14}C-labelled saccharin was synthesized for metabolic studies [29].

A thiophene analogue of saccharin (3-oxo-2,3-dihydrothieno-[3,4-*d*]isothiazole-1,1-dioxide) was synthesized in 1980

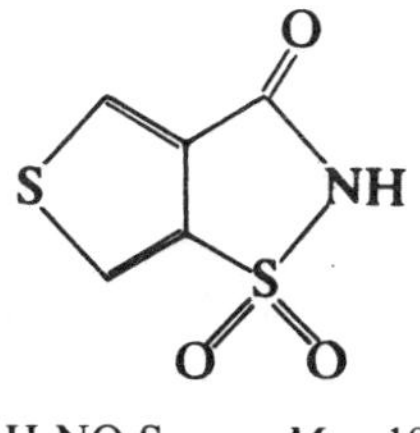

$C_5H_3NO_3S_2$, $M = 189{,}1$

and it was found to be a thousand times sweeter than the reference sucrose, twice as sweet as saccharin itself, but free of its characteristic secondary taste [22]. The starting materials for the synthesis were the methyl esters of acrylic and thioglycolic acids (Scheme 3.4).

A closely related [2,3-*d*] isomer was also synthesized as shown in Scheme 3.5.

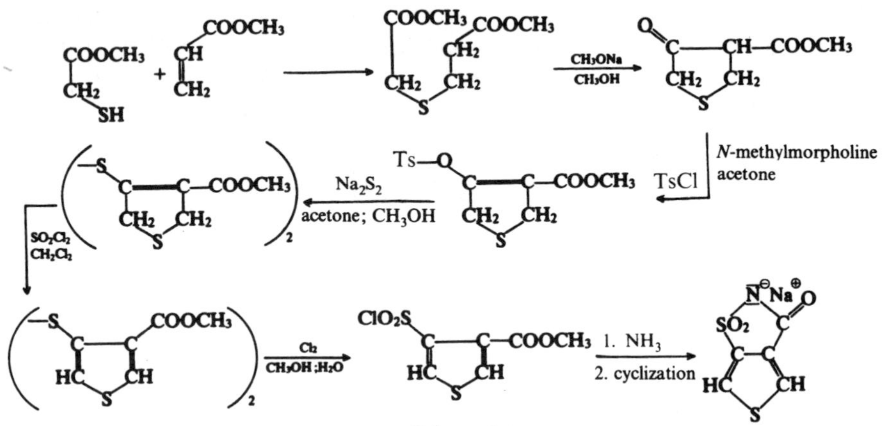

Scheme 3.4

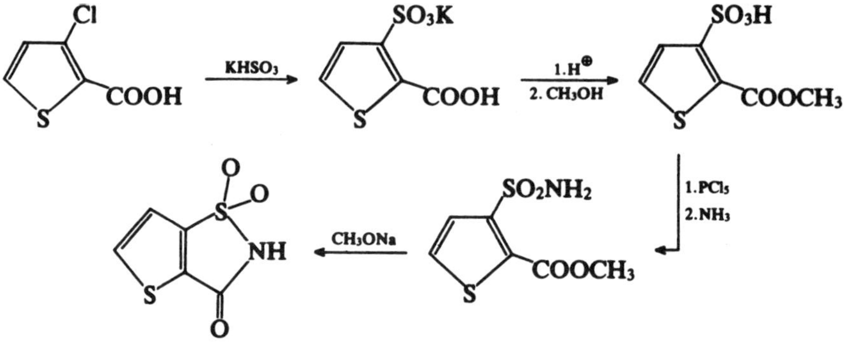

Scheme 3.5

The next Scheme 3.6 presents the chemistry involved in Wannagat's synthesis of sila-saccharin [26].

Contrary to expectations, neither sila-saccharin, nor its lithium or sodium salts were found to be sweet.

Because of the large acidity of the imidic hydrogen, many *N*-substituted derivatives of saccharin were successfully prepared utilizing the phase transfer catalysis (Scheme 3.7) [24].

The *N*-substituted derivatives, when treated with nucleophiles (e.g. hydrazine, hydroxide anion), open the five-membered isothiazole ring:

R = H, Na, alkyl, aryl
Nu = $OH^{\ominus}$, hydrazine

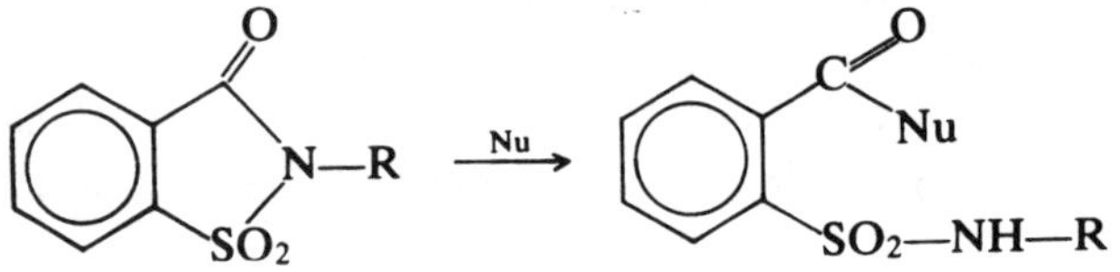

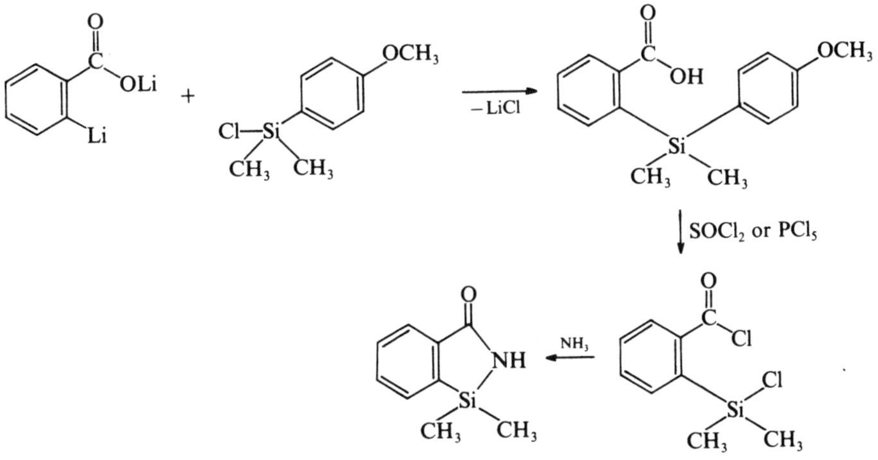

$-LiCl$

$SOCl_2$ or PCl_5

NH_3

sila-saccharin

Scheme 3.6

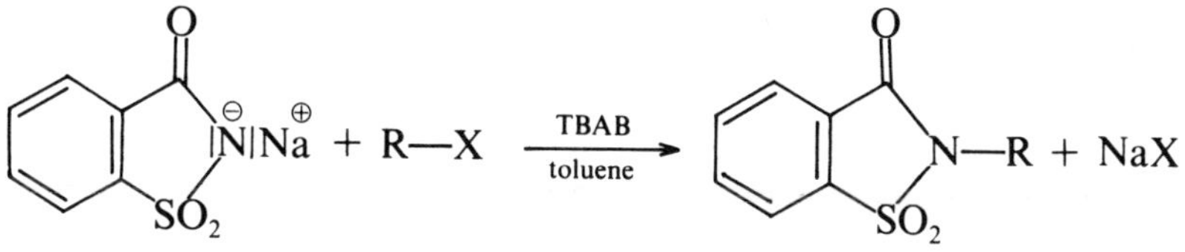

TBAB

toluene

R = alkyl, acyl
TBAB = tetra-*n*-butylammonium bromide

Scheme 3.7

Saccharin is marketed either as a free acid or in the form of its more soluble salts under many brand names, such as Azucaretas, Bayer Süßstoff, Dulcetas, Dulsacar, Edulcor, Garantose, Glucid, Gluside, Hermesetas, Kandiset, Kristalose, Natrium Benzsulfimidicum, Oda, Sacarina, Saccharina, Saccharinol, Sacharin T 1000, Sacharinum Solubile, Saxin, Skun-Suc, Sucrine, Sucrette, Sucromat, Sun-Suc, Sykose, Zaharina, Zuckerin [12, 14, 25].

3.1.2 Analytical determination of saccharin

Within the limits of the present analytical methods, saccharin produced by the Fahlberg–Remsden method (Scheme 3.1) contains 31 impurities. Phthalic anhydride method (Scheme 3.2) yields 23 impurities [12]. Some of the impurities, identified by chromatographic methods, are listed by

Fabiani [14]. Nelson [27] confirmed the presence of 2- and 4-sulphamoylbenzoic acids; and saccharin prepared by Maumee's process contains also aminosaccharins [15].

Analytical determination of saccharin in drugs and soft drinks is carried out by a combination of thin-layer chromatography, extraction and UV spectrophotometry [35, 36]. The application of UV spectrophotometry for the assay of derivatized sweeteners in cola beverages is described by Stolzberg [37]. Analytical determination of saccharin in non-alcoholic drinks, juices and cocktails can be also carried out by the reverse phase high-pressure liquid chromatography [39–42], gradient liquid chromatography methods which enable simultaneous determination of sweeteners, preservatives and colouring agents [43] and ion-pair chromatography [44] which is applicable to the analysis of foodstuffs.

Saccharin in food was also determined by the application of differential pulse polarography [45].

Saccharin anion can migrate in an electric field and therefore can be separated from other interfering substances, which are found in food and soft drinks, by capillary isotachophoresis [47].

Other analytical methods which are suitable for saccharin detection are the non-aqueous titrations which utilize Sudan Blue GN and Sudan Green 4B as indicators [48].

3.1.3 Structure–sweetness relationship

Three generalizations could be derived from the study of a set of more than 80 saccharin derivatives [21]:

(a) exchange of the imidic proton for cations other than Na^+, NH_4^+ or Ca^{2+} decreases the sweetness. All *N*-alkyl, cycloalkyl and *N*-heterocyclic substituted derivatives are tasteless
(b) electron withdrawing substituents attached to the aromatic nucleus induce bitterness
(c) electron donating substituents, such as NH_2, CH_3 or C_2H_5O, increase the sweetness

A change in the NH functional group may diminish the sweetness whilst the substitution of sulphur by selenium removes the sweetness completely. Furthermore, structural changes in the saccharin molecule manifest themselves in a range of other sensory (taste) properties, which are summarized in Scheme 3.8.

Rader and co-workers [34] published a detailed study of the taste*

* The taste is defined as a sensation perceived solely through the mouth cavity, in contrast to deliciousness which is the quality affected by the combined sensation of taste,

1,2-benzisothiazolin-1,1-dioxide

3-amino-benz[*d*]isothiazole-1,1-dioxide

benz[*d*]isothiazole-1,1-dioxide

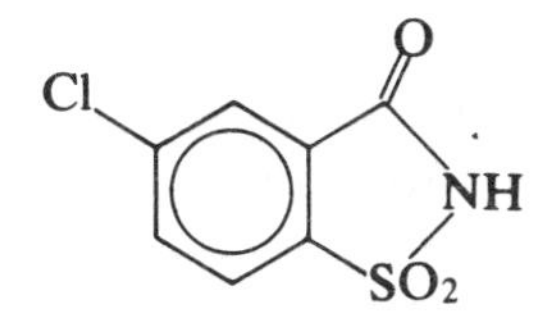

5-chlorosaccharin

Scheme 3.8

of saccharin. Chay and Hayert [30] assayed the sweetness of refreshing non-alcoholic beverages by the triangle method. According to Redlinger and Setser [31], who evaluated a range of sweeteners, flavour did not affect the initial, maximal or residual sweetness but the lack of sweetness in the aftertaste was less intense in lemon and plain unbaked cookies than in vanilla-flavoured unbaked cookies.

3.1.4 Properties and utilization of saccharin

Although saccharin has been used continuously since its discovery, its application was never very extensive as might have been assumed from the length of its use. This is due to its sensory properties which are not unambiguously positive and are marked by the persistent bitter aftertaste, thus limiting its use in the food industry. The remedy to this problem is usually the use of combined sweeteners, either with a neutral (tasteless) carrier such as gelatin, or in combination with other sweeteners (see p. 214). A commonly used mixture is a pellet-form combination of the sodium salt and sodium bicarbonate (1 : 4) which releases carbon dioxide upon dissolution in water.

The potential health hazards connected with the use of saccharin have been for some time a hotly debated topic. Nevertheless, the retention time of saccharin in the human body is rather short; it quickly passes through the digestive organs and up to 98 % of it leaves the body in urine [14].

Likewise, its acute toxicity is small; LD_{50} by intraperitoneal application in mice was 17.5 g/kg, intravenal application to dog gave 2.5 g/kg

smell and physical consistency of the food (bulk, structure, mouth-feel, viscosity) as well as by the visual feedback [38].

and per os application to rabbit 8 g/kg. Similarly, chronic toxicity poses no danger. Extended studies [25] have shown no mutagenicity; the first case of mutagenicity is reported in reference [49]. Teratogenic and cytotoxic activity of saccharin is acceptably low [25]. Epidemiological studies of Arnold and co-workers [12] have revealed that there is a certain relationship between the use of saccharin and urinary bladder cancer incidence in rats. Their findings were corroborated in 1978 by Hicks' group [53] who studied co-carcinogenesis of saccharin and cyclamates in experiments where the urinary bladder cancer in rats was induced by application of nitrosourea. Similar results were obtained with carcinogens such as *N*-[4-(5-nitro-2-furyl)-2-thiazolyl]formamide [50] and *N*-butyl-*N*-(4-hydroxybutyl)nitrosoamine [51].

Saccharin, as an acidic cyclic sulphonamide, apparently passes through the gastrointestinal tract without being metabolized, and quickly reaches urine. The observed incidence of urine bladder tumours in male rats which received high saccharin diet from birth cannot be therefore simply explained by the accumulation of the sweetener in the bladder [79].

Test animals failed to show any substantial carcinogenic effect even when high doses of saccharin were administered: however, animals which were pretreated with a low (non-carcinogenic) dose of *N*-methyl nitrosourea quickly developed urine bladder carcinoma in as many as 60% of the cases. This discovery indicates that the study of carcinogenicity of the suspect substance alone does not suffice to assess the potential risks associated with its use. Thus we may conclude that saccharin should not be used daily throughout the lifetime.

In parallel with the unequivocal discussion on the possible detrimental effects of saccharin on human health, the limits of its applications and usage are also not clearly defined. Thus the pharmaceutical industry makes frequent use of it as a taste rectifier, as an additive to tooth paste and mouth lotions; additionally, saccharin has found its way into chewing gums and special foods for people suffering from diabetes mellitus. It was used routinely in the last world war as a substitute for sucrose. It is also used as a favoured sweetener by people with an inclination to fat deposition, diabetes patients and people suffering from gastrointestinal disorders.

At present, the use of saccharin is permitted by countries affiliated to the World Health Organization, and is limited by the 5 mg/kg of product weight content (15 mg/kg for diabetics).

The recommended daily dose is 15 mg per 1 kg of body weight, or simply about 1 g/day.

3.2 DERIVATIVES OF AMINO-*N*-SULPHAMIC ACID — CYCLAMATES (SULPHAMATES)

Cyclamates — a new family of sweeteners — were discovered accidentally, just as saccharin was, in the course of a study of antipyretic properties of *N*-alkylated derivatives of aminosulphamic acid conducted at the Illinois State University in 1937. M. Sveda, one of the discoverers, found his cigarette tasted sweet. The derivatives of aminosulphamic acid, including those of cyclohexylsulphamic acid as well as their salts, were patented in 1940 as novel low-energy sweeteners. The first paper on the topic by Sveda was published much later in 1944 [54]. The commercial exploitation of the salts of *N*-cyclohexylsulphamic acid started in the US with the sodium salt in 1950 and the calcium salt in 1954.

N-Cyclohexylsulphamic acid is a white, crystalline sweet compound (without any other additional flavours), with a melting point of 169–170 °C. It dissolves readily in water (1 g/13 ml) as well as in polar

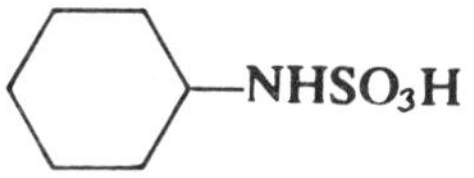

—$NHSO_3H$

$C_6H_{13}NO_3S$, $M = 179.24$

organic solvents (1 g/25 ml ethanol, 9 ml glycerol or 25 ml 1,2-propanediol it is also very soluble in ethyl acetate and acetone), but is only sparingly soluble in chloroform and benzene and insoluble in hexane. A 10% aqueous solution has a pH in the range of 0.8–1.6. In applications that require an addition of basic (alkaline) constituents, such aqueous solutions are stable over a range of pH values from 2 to 10.

The sodium salt of the *N*-cyclohexylsulphamic acid resembles the parent compound: it is a white crystalline powder with a sweet taste and is thermostable (decomposes above 260 °C). Although its solubility in water is still good (1 g/5 ml or 24 ml 1,2-propanediol), it is practically insoluble in ethanol (75% aqueous ethanol gives only 1 % solution at 25 °C), benzene, chloroform, acetone or ether. Its aqueous solutions are neutral. The sweetness of the sodium salt of *N*-cyclohexylsulphamic acid is 30 times that of sucrose. The sweetness of the solution fades with increasing concentration of the sweetener. The optimum is reached at concentration as low as 1% — at higher concentrations the sweetness levels off and the sensation becomes somewhat diluted.

The sweet taste of the calcium salt, like that of the sodium salt, is free from disturbing additional tastes. It crystallizes with 2 molecules of water which can be removed by heating at 80 °C, dry salt decomposes at 260 °C. One gram of the calcium salt dissolves in 4 ml of water, 60 ml of ethanol or in 1.5 ml of 1,2-propanediol. It is insoluble in benzene, chloroform and ether and its aqueous solutions are neutral. Its sweetness matches that of the sodium salt. The solubility of both salts can be improved by mixing them in the ratio of 1 : 0.5 to 1 : 1.2.

Other salts (K, Mg) are sweet as well, but have never reached the commercial stage of exploitation [55].

3.2.1 Preparation of cyclamates

Sveda's original preparation of *N*-cyclohexylsulphamic acid involved an addition of chlorosulphonic acid (1 mole) to a solution of cyclohexylamine (3 moles) in an inert solvent (chloroform, benzene or cyclohexane) at 0–20 °C, giving cyclohexylammonium *N*-cyclohexylsulphamate and cyclohexylammonium chloride (Scheme 3.9).

$$3\ C_6H_{11}\text{-}NH_2 + ClSO_2OH \xrightarrow[5°C]{CCl_4} C_6H_{11}\text{-}NHSO_3^{\ominus}\ H_3\overset{\oplus}{N}\text{-}C_6H_{11} + C_6H_{11}\text{-}\overset{\oplus}{N}H_3\ Cl^{\ominus}$$

$$\downarrow NaOH$$

$$C_6H_{11}\text{-}NHSO_3^{\ominus}\ Na^{\oplus} + C_6H_{11}\text{-}NH_2 + H_2O$$

Scheme 3.9

Subsequent treatment with sodium hydroxide yields the sodium *N*-cyclohexylsulphamate and cyclohexylamine (2 moles) which is distilled out. The salt is then purified by crystallization from water. Other commercial procedures utilize also aminosulphonic acid as well as sulphur trioxide [56]. Tertiary amine such as triethylamine or trimethylamine can be used in place of two moles (out of the three required) of the starting amine. Some of the processes mentioned above are shown in Scheme 3.10.

The key step in all cases above is the formation of cyclohexylammonium *N*-cyclohexylsulphamate which is subsequently decomposed by sodium hydroxide or calcium hydroxide into cyclohexylamine and the sodium or calcium salt of *N*-cyclohexylsulphamic acid, respectively. Patent literature [57] claims that cyclohexylammonium *N*-cyclohexylsulphamate is also formed in 78 % yield when cyclohexylamine is reacted with sodium disulphite at 25 °C.

Alternative synthetic procedures describe the use of cyclohexylhydroxylamine, which on treatment with sulphur dioxide gives *N*-

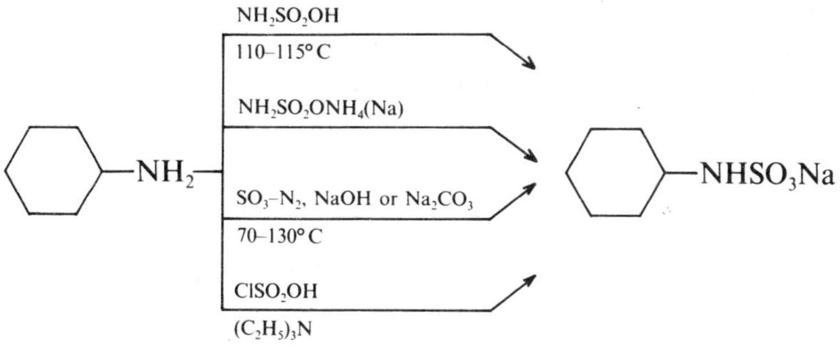

NH_2SO_2OH, 110–115°C; $NH_2SO_2ONH_4(Na)$; SO_3–N_2, NaOH or Na_2CO_3, 70–130°C; $ClSO_2OH$, $(C_2H_5)_3N$

Scheme 3.10

cyclohexylsulphamic acid, or yields directly the sodium salt when reacted with sodium bisulphite. Yet another synthetic scheme utilizes the catalytic reduction of aromatic sulphamates [23].

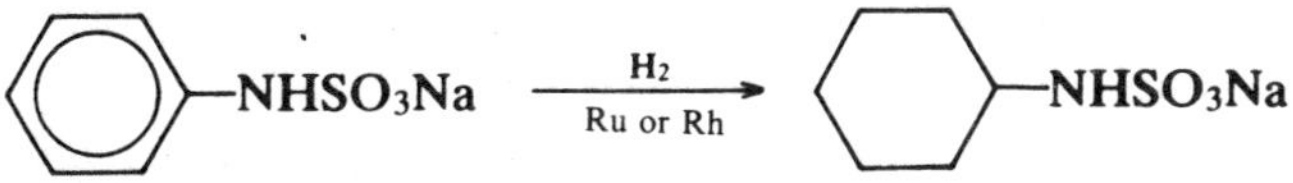

$C_6H_5NHSO_3Na \xrightarrow[\text{Ru or Rh}]{H_2} C_6H_{11}NHSO_3Na$

In addition to the work of Audrieth and Sveda [54], cyclamates have been also studied by Nofre and Pautet [58–61] and Unterhalt and Böschemeyer [62–64]. Some of their works are summarized in Scheme 3.11.

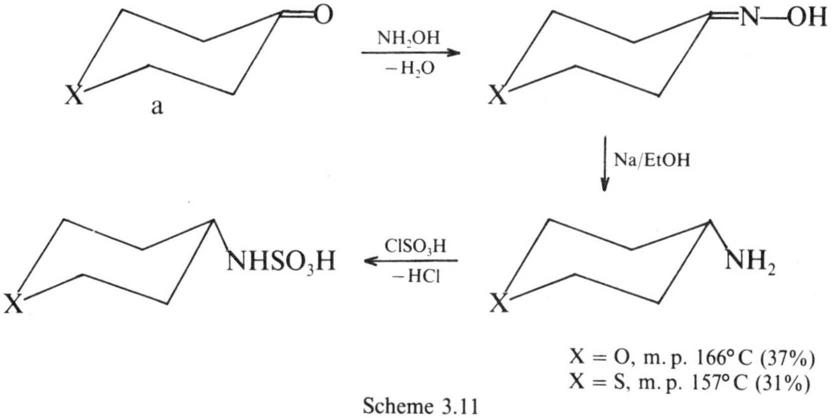

a; NH_2OH, $-H_2O$; Na/EtOH; $ClSO_3H$, $-HCl$

X = O, m.p. 166°C (37%)
X = S, m.p. 157°C (31%)

Scheme 3.11

The starting ketone (a) was prepared by Dieckmann's condensation [64].

The original synthesis of the parent cyclamate is based on the reaction of cyclohexyl isocyanate with concentrated sulphuric acid. When the reaction is completed, the reaction mixture is poured onto an ice–water mixture which contains calcium oxide and calcium hydroxide. Calcium sulphate precipitates and the product (as the calcium salt) is purified by crystallization (Scheme 3.12).

$$C_6H_{11}{-}NCO + H_2SO_4 \longrightarrow C_6H_{11}{-}NHSO_2OH + CO_2$$

$$\downarrow CaO + Ca(OH)_2,\ -CaSO_4$$

$$(C_6H_{11}{-}NHSO_2O)_2Ca$$

Scheme 3.12

Cyclamates prepared by the synthetic procedures described in the patent literature [65, 66] must be rigorously purified by recrystallization with the addition of charcoal, diatomaceous earth, by extraction with organic solvents or by filtration through ion exchange resins. The literature data [56] state that these procedures give, quite incredibly, yields from 75 to 95 %.

Sulphamates are chemically rather inert; one of the few known reactions is their decomposition with nitrous acid when the N–S bond is cleaved. The reaction has been utilized for the analytical determination of cyclamates [55].

$$C_6H_{11}{-}NHSO_3^{\ominus} + HONO \longrightarrow C_6H_{10} + HSO_4^{\ominus} + N_2\uparrow + H_2O$$

The most commonly used cyclamates (Na, Ca) have been marketed under the following brand names: Assugrin, Azucrona, Cyclan, Cyclarin, Cyclamat, Ibiosuc, Makeryl, Natrena, Nectaryl, Satomin, Spolarin, Sucaryl, Succrosa, Sussin, Sweeta, Sweet-Ten [2, 25, 71].

3.2.2 Structure–taste relationship for sulphamate sweeteners

Audrieth and Sveda were the first researchers who tried to establish which part of the cyclamate moiety is responsible for eliciting the sweet taste response. Their structure–activity study revealed that the presence of the cyclohexane ring stimulated sweetness. Nofre and Pautet established in 1975 that *n*-butylcyclamate was 50 times sweeter than sucrose. However, it appears that the hydrogen atom in the $NHSO_3Na$

moiety is absolutely essential for sweetness, since all substituted derivatives (methyl, ethyl, cyclohexyl, phenyl) were tasteless.

Extensive studies of Benson and Spillane [67] helped to establish the structure–taste relationship for cyclamates. They suggest that these relationships are best categorized under the headings: size of the reduced (saturated) ring, changes in the sulphamate function, substitution of hydrogen and carbon of the alicyclic ring and opening of the ring. Their results indicated that:

Sulphamates with medium ring are sweet; those with small rings are tasteless.

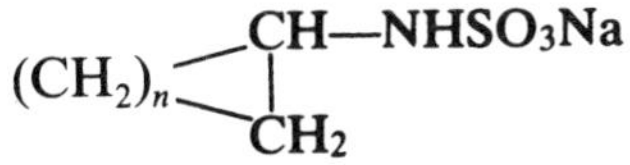

$(CH_2)_n$ ring: $CH-NHSO_3Na$ / CH_2

$n = 3–7$ — sweet, $n = 1, 2$ 10 — tasteless

Substitution of the ring hydrogen for either methyl, ethyl or two methyl groups does not affect the sweetness [23], e.g.

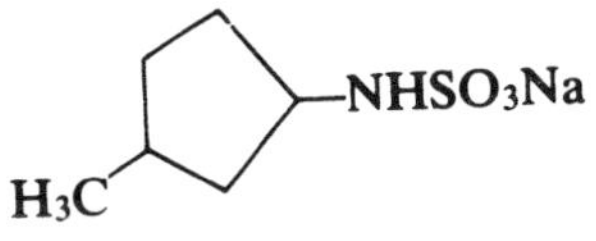

H_3C–cyclopentyl–$NHSO_3Na$ 30 times sweeter than sucrose

An insertion of a methylene group between the sulphamate group and the cycloalkyl residue may in some cases reduce the sweetness [59], for example

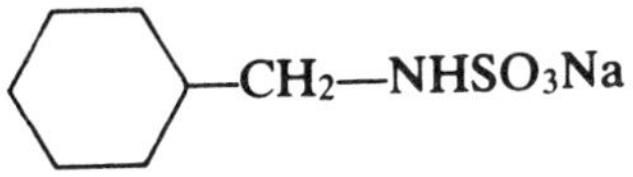

cyclohexyl–CH_2—$NHSO_3Na$ cyclopentyl–CH_2—$NHSO_3Na$

tasteless 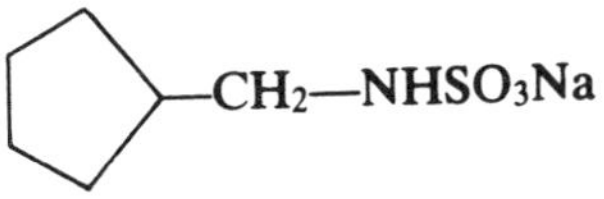 sweeter than *N*-cyclohexylsulphamate

N-Alkylation of the sodium salt of *N*-cyclohexylsulphamate results in the loss of sweet taste, for instance

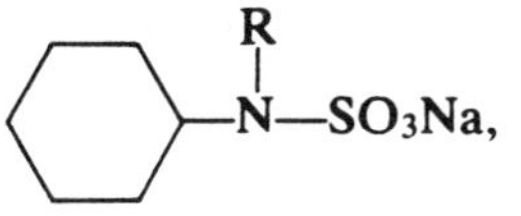

cyclohexyl–N(R)—SO_3Na,

tasteless

$R = C_1–C_3$, cycloalkyl group

Similarly, the replacement of the 4-methylene group in the cyclohexane by an oxygen atom removes the sweetness, but the replacement by a sulphur atom or a sulphonyl group retains it.

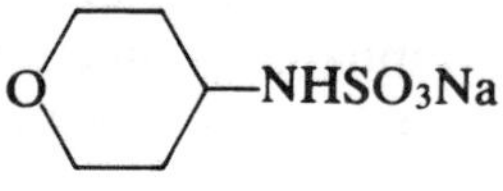

$-NHSO_3Na$

tasteless

$-N(R)-SO_3Na$

$-N(R)-SO_3Na$

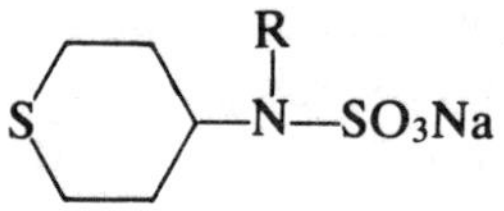

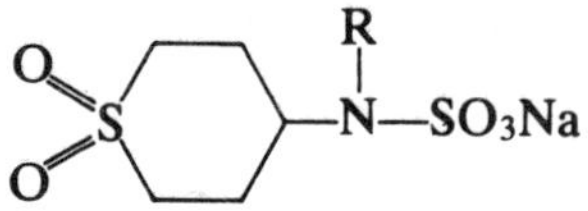

sweet

R = ethyl, *n*-propyl, *n*-butyl

The sweet taste is also lost in *N*-piperidinyl and *N*-morpholinyl derivatives. Thus it appears that the manifestation of sweetness requires the presence of hydrogen atom at C_1 [67] as can be shown

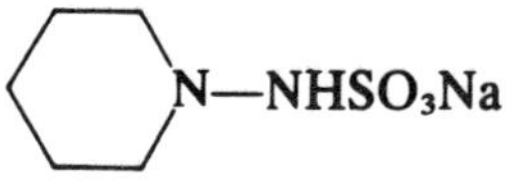

$N-NHSO_3Na$

tasteless

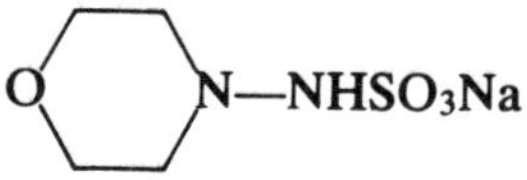

$N-NHSO_3Na$

tasteless

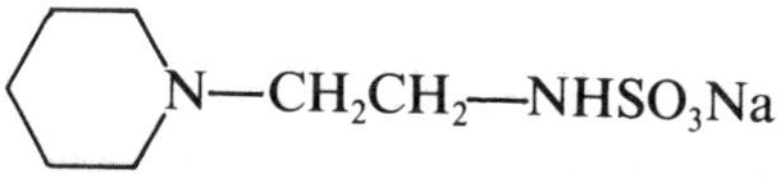

$N-CH_2CH_2-NHSO_3Na$

bitter

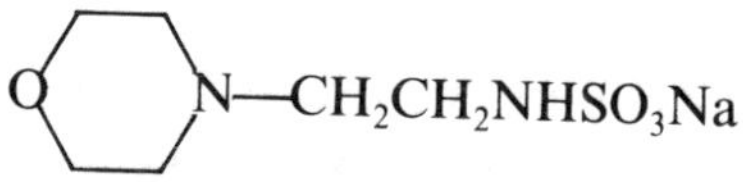

$N-CH_2CH_2NHSO_3Na$

tasteless [68]

but the following two examples:

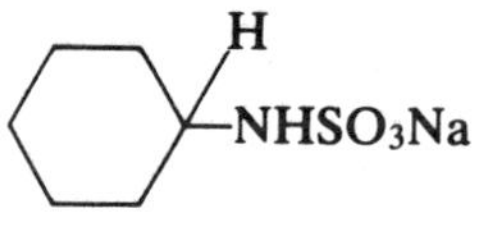

H, $-NHSO_3Na$

sweet

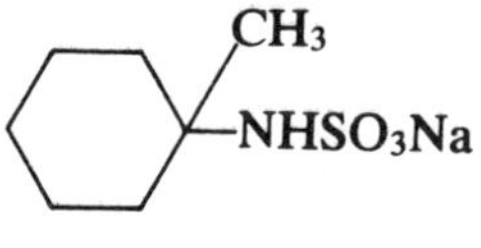

CH_3, $-NHSO_3Na$

tasteless

The intensity of the sweetness depends on the length and the nature of the group attached to the nitrogen atom as is illustrated for the straight and branched chain alkyl radicals

$CH_3-(CH_2)_n-CH_2NHSO_3Na$

$n = 1, 2$ — sweet

$n = 4, 6, 13, 16$ — tasteless

$(H_3C)_2CH-(CH_2)_n-NHSO_3Na$

$n = 1, 2$ — sweet

Pautet and Dandon [69] reported a group of 2-substituted compounds derived from ethyl sulphamate and found that only the methylthio derivatives were sweet

$$R–CH_2–CH_2–NHSO_3Na,$$

where R = OH, CH_3O, $N(CH_3)_2$, COONa are tasteless, but R = CH_3S is sweet.

Sensory properties of the synthesized *N*-(1-adamantyl) sulphamates [67] confirmed again the hypothesis that the presence of

$$\rangle CH–\underset{\displaystyle R}{\underset{|}{N}}–SO_3^{\ominus}$$

is essential for sweetness. The replacement of the SO_3^- group with other polar groups, such as SO_2Cl [68], SO_2R, COOR, NO_2, $COCOO^-$, $CH_2SO_3^-$, $COSO_3^-$ [70] leads to the formation of non-sweet compounds.

The authors [70] also point out the importance of the conformation of the polar group, the optimal torsion angle between the N–H and S–O bonds is about 60 degrees, as in the synclinal conformation (Fig. 3.2).

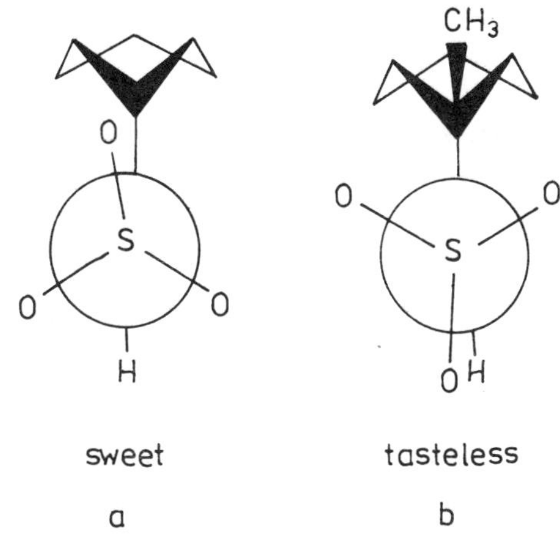

Fig. 3.2 — Newman projection along the N—S bond of a — *N*-cyclohexylsulphamate (torsional angle 60°, synclinal conformation); b — *N*-(1-methylcyclohexyl)sulphamate (torsional angle 0°, synperiplanar conformation) [61].

Spillane and McGinckey assessed the structure–activity relationship of cyclamates using the space-filling atomic models of Corey–Pauling–Koltun (CPK) [98]. Their study showed that the best $NHSO_3$ group conformation for sweet taste stimulation involved an angle of 60 degrees between the N–H and S–O bonds. The need to maintain this optimum torsional angle explains why the substitution of an α-hydrogen by an alkyl group at C_1 of an alicyclic ring or an aliphatic chain destroys the sweetness.

3.2.3 Determination and utilization of sulphamates

The analytical methods published up to 1967 were reviewed by Richardson [72]. Two methods feature prominently in the quantitative determination of the cyclohexylsulphamic acid; one is based upon the elimination of cyclohexene which is analysed by gas chromatography, the other utilizes the reaction with chlorine in aqueous media yielding *N,N*-dichlorocyclohexylamine. The amine is separated from the reaction mixture by extraction with cyclohexane and its amount determined by its UV absorption at 314 nm [73, 74].

$$C_6H_{11}\text{—}NHSO_3H + 2\,Cl_2 + H_2O \xrightarrow{H^{\oplus}} C_6H_{11}\text{—}NCl_2 + 2\,HCl + H_2SO_4$$

The quantitative determination of sodium cyclamate in soft drinks makes use of several techniques; the assay combines extraction, thin-layer chromatography and UV spectroscopy aided by a colouring agent [35, 36].

Thin-layer chromatography is performed on polyamide and utilizes various solvent mixture as eluents such as xylene–*n*-propanol–formic acid in the ratio 5 : 5 : 1 or benzene–formic acid–ethyl acetate 5 : 2 : 10 [41, 75].

Cyclamates in food [46] or fruit juices [47] were also determined by capillary isotachophoresis or reverse phase liquid chromatography. The other advanced chromatographic techniques in use are high-pressure liquid chromatography (HPLC) with indirect photometric detection [77] and high-performance ion chromatography [78].

By analogy with the saccharin methodology, cyclamates can also be determined by titrations in non-aqueous solutions [48].

The worldwide use of cyclamates has generated sufficient momentum to stimulate extensive research efforts aimed at proving that the health hazards are low. As a consequence, the ban on the use of cyclamates in several countries has been lifted again.

In comparison with saccharin, cyclamates elicit not only a more natural sweet sensation, their physico-chemical properties are also less objectionable than those of saccharin. They are rather stable on heating during a baking process or being heated in solution, even at higher temperatures, their shelf-life is satisfactorily long, they are not hygroscopic and yet still very soluble in water. They are stable in solutions over a large pH range (2–10) [71].

Both the sodium and calcium salts serve as a taste rectifier as they successfully eliminate bitter taste even at low dosages. They combine well with fruit flavour additives. All the properties mentioned above have made them the favourites of the pharmaceutical industry, both in human and veterinary medicine, and they have become really indispensable in the production of preparations for diabetics. They are also used for the direct taste correction of acetylsalicylic acid (aspirin), for many orally administered antibiotics, sulphonamides, antihistamines, barbiturates and many others [71]. Sulphamates do not stimulate the growth of microflora and moulds.

The greatest world producers and consumers of cyclamates are the US and Japan. The main bulk of the sweetener, mainly as the calcium salt, is used in soft drinks; the sodium salt is less suitable as some people are constrained to a low sodium diet. Nevertheless, even the calcium salt is not recommended for young children and pregnant women.

There is another indirect but very important benefit to be gained from the use of cyclamates. They help to control the body weight and the energy load of human organism. Cyclamates are also more acceptable to diabetics when they are often combined with saccharin or neohesperidine dihydrochalcone; in the latter case they do exhibit synergy.

Cyclamates are currently used on a large scale in the food industry. They are a component of soft drink concentrates, fruit juices, preserved fruit, jams, pickles, glazes, cream fillings, jellies, fruit fillings, pastries, confections, as well as many meals in the form of desserts and puddings, and in coffee and tea. Cyclamates have been recently marketed in an effervescent form as tablets [101] and now even the sodium salt of saccharin and aspartame are being offered in this popular form. The tablets release the sweetener instantaneously even at low humidities. Two methods of the sweetener incorporation in the tablet have been devised and the resulting product compared and tested with respect to the galenic and physico-chemical tests (IR and X-ray diffraction).

The parent *N*-cyclohexylsulphamic acid has a pleasant sweet taste that complements well the sour lemon flavour. The unpleasant taste of some per os administered drugs can be improved by converting them into the sulphamate salts if they have a basic functional group in the molecular structure. This simple chemistry improves, in general, the solubility of the drug, also or at least simplifies its application.

The use of cyclamates must be clearly declared on the packaging of all products that contain them. Although the administration of cyclamates in large doses (5–10 g per day) can cause lethargy and diarrhoea, specially without a parallel adequate food intake, their animal toxicity is quite low. The LD_{50} values for mice per os are 10–17 g/kg for the free acid and 7.2 g/kg for the calcium salt; the dose for rats (per os application) is 15.3 g/kg of the sodium salt. It was shown that cyclamates have a teratogenic effect [25] when used over a long time. The present recommended daily dose is 11 mg/kg of body weight.

No adverse effects on liver or kidneys could be detected in humans. The cyclamate metabolites leave the body very quickly and almost completely in the urine. However, a small part was found to persist in the gastrointestinal tract (0.1–0.9 %) where it was metabolized into the strongly toxic cyclohexylamine, or even into the carcinogenic dicyclohexylamine [76]. When a ^{14}C-labelled sodium cyclamate was administered to rats, rabbits and guinea pigs, the correspondingly labelled cyclohexylamine was found in their urine [79]. However, the length of the metabolic pathway varies widely not only in different animal species but also within the species.

Hicks and co-workers [53] concluded that the calcium salt of cyclohexylsulphamate stimulates, in the same way as saccharin does, the growth of the urine bladder tumours in rats if the animals were pretreated with a subthreshold dose of methyl nitrosourea. In contrast, no carcinogenic effect was elicited, even when a large dose of cyclamate was administered in food, but in the absence of any carcinogen. The bladder carcinoma developed in 60 % of cases only if the carcinogen was administered in advance [52].

3.3 OXATHIAZINONE DIOXIDES

Clauss and Jensen [80] described in 1973 a new family of synthetic sweeteners — substituted 3,4-dihydro-1,2,3-oxathiazin-4-one-2,2-dioxides; the 6-methyl derivatives having a specially strong sweet taste (Table 3.3).

Oxathiazinone dioxides are chemically and thermally stable crystalline compounds, soluble in water and organic solvents (methanol, ethanol, glycerol) as well. They dissociate in aqueous solutions and form neutral salts with melting points above 200 °C.

This group of synthetic sweeteners is usually prepared by the reaction of fluoro-, chloro- or arylsulphonyl isocyanates with carbonyl compounds or alkynes.

Thus the reaction of 2-butyne with fluorosulphonyl isocyanate afforded a derivative of uracil (a), the structure of which was confirmed by an X-ray analysis [81]. The hydrolysis of the uracil gave an amide which was then cyclized in basic media to the oxathiazinone dioxide in 75% yield (Scheme 3.13).

Table 3.3 — Relative sweetness of oxathiazinone dioxides (4% sucrose solution)

R^1	R^2	X	Relative sweetness	References
H	H	H	10, 20	[19, 23, 80]
H	H	Na	10	[19, 80]
H	Me	Na	130	[80]
H	Me	K	130, 200[a]	[28, 30–32, 34]
			180	[80, 82, 83, 85, 99, 100]
H	Me	Ca	130	[80]
Me	H	Na	20	[80]
Me	Me	Na	130	[80]
H	Et	Na	150	[80]
Et	H	Na	20	[80]
Me	Et	Na	130	[80]
Et	Me	Na	250	[80]
H	*n*-Bu	Na	30	[80]
n-Pr	Me	Na	30	[80]
Et	*n*-Pr	Na/K	70	[80]
i-Pr	Me	Na	30, 50	[20, 80]
Et	*n*-Bu	Na	70	[80]
H	CH_2Cl	K	150	[85]
H	CH_2Br	K	150	[85]
H	CH_2OCH_3	K	50	[85]
H	$CH_2OC_2H_5$	K	bitter	[20]
H	CH_2OH	K	50	[85]
H	$CH_2—N_3$	K	150	[85]
Cl	Me	K	200	[20]

[a] 3% solution of sucrose.

$CH_3—C≡C—CH_3$ + 2 $O=C=N—SO_2F$ → (a) $\xrightarrow{H_2O}$ → NaOH, $-CO_2↑$ →

75%, m.p. 108.5 °C

Scheme 3.13

Analogous products have been synthesized starting with 1-butyne and 1-hexyne; however, this synthetic approach is not viable on a commercial scale.

More satisfactory results are achieved by a method based on the reaction of fluoro- or chlorosulphonyl isocyanate with carbonyl compounds, for instance 2-butanone, acetone, 2,4-pentanedione or with *t*-butyl acetoacetate (Scheme 3.14).

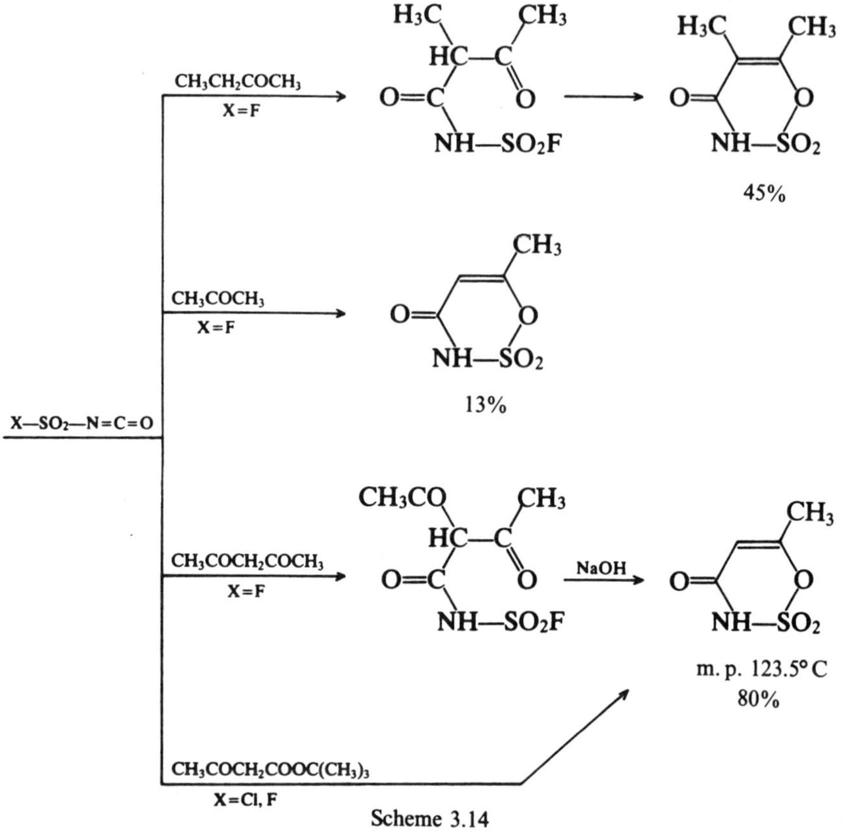

$X{-}SO_2{-}N{=}C{=}O$

$CH_3CH_2COCH_3$, X=F → H_3C, CH_3, HC—C, O=C, O, $NH{-}SO_2F$ → H_3C, CH_3, O=, O, $NH{-}SO_2$ 45%

CH_3COCH_3, X=F → CH_3, O=, O, $NH{-}SO_2$ 13%

$CH_3COCH_2COCH_3$, X=F → CH_3CO, CH_3, HC—C, O=C, O, $NH{-}SO_2F$ —NaOH→ CH_3, O=, O, $NH{-}SO_2$ m. p. 123.5° C 80%

$CH_3COCH_2COOC(CH_3)_3$, X=Cl, F

Scheme 3.14

The reaction with the last compound, *t*-butyl acetoacetate, proceeds in ether at 0–20 °C. The intermediate (1) on heating to 40–70 °C gives off carbon dioxide and 2-methylpropene together with the *N*-X-sulphonamide of the 3-oxobutyric acid. The sulphonamide is finally cyclized in alkaline conditions to the 6-methyl derivative (Scheme 3.15).

The industrial process for the above compound utilizes the fluorosulphonyl isocyanate and the cyclization is accomplished by the action of potassium hydroxide, leading directly to the potassium salt which is marketed as Acetosulfam [82], Acesulfam K [83] or Sunett [84]. Recrys-

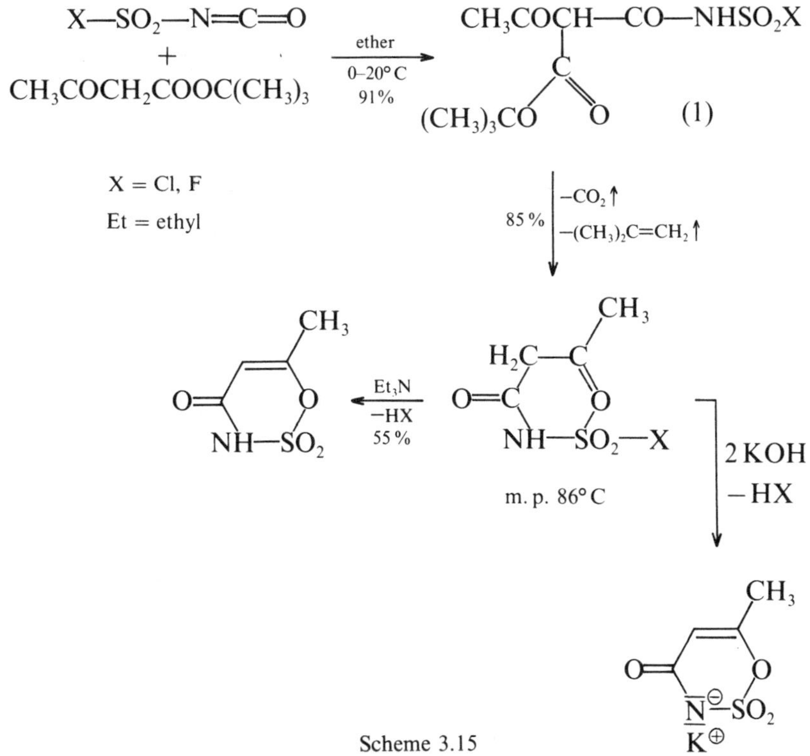
X—SO2—N=C=O
+
CH3COCH2COOC(CH3)3
ether
0–20° C
91%
CH3COCH—CO—NHSO2X
(CH3)3CO
O
(1)
X = Cl, F
Et = ethyl
85%
−CO2↑
−(CH3)2C=CH2↑
CH3
H2C
O
NH—SO2—X
m. p. 86° C
Et3N
−HX
55%
NH—SO2
2 KOH
−HX
K⊕

Scheme 3.15

tallization from water yields a product with 99 % purity (water solubility at 20 °C is 27 g/100 ml, 130 g/100 ml at 100 °C). The tentative mechanism of the cyclization is given in Scheme 3.16.

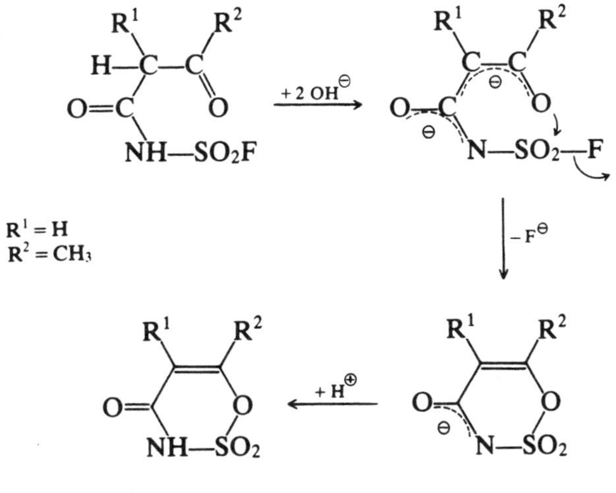
R1
R2
H
O
NH—SO2F
+ 2 OH⊖
N—SO2—F
R1 = H
R2 = CH3
− F⊖
+ H⊕
NH—SO2
N—SO2

Scheme 3.16

When the reaction is carried out with 4-substituted *t*-butyl-3-oxo-butanoates, it affords 6-halomethyl, 6-alkoxymethyl, 6-acetomethyl and 6-azidomethyl derivatives in very good yields. Their structures were confirmed by ^{1}H NMR spectroscopy [85].

The salts of the oxathiazinone dioxides dissolve readily in water and they can be isolated from aqueous solution as the free acids by acidification, followed by an extraction with an organic solvent such as ethyl acetate. The benzo derivative of oxathiazinone dioxide (4-oxo-1,2,3-benzoxathiazine-2,2-dioxide) was prepared from 2-benzyloxybenzoic acid (Scheme 3.17). It is a hundred times sweeter than sucrose [23].

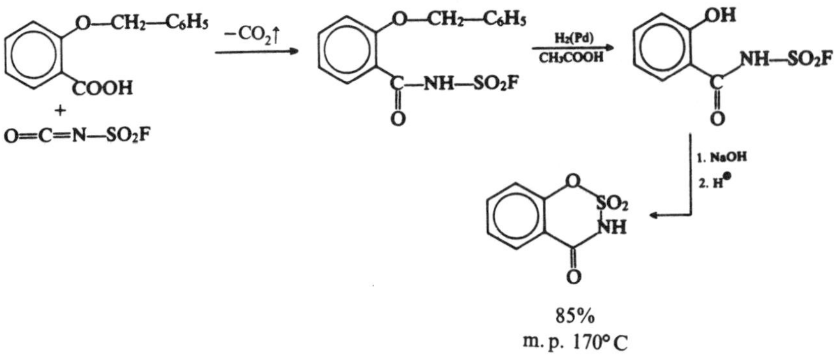

O—CH_2—C_6H_5 / COOH + O=C=N—SO_2F → ($-CO_2\uparrow$) O—CH_2—C_6H_5 / C(=O)—NH—SO_2F → (H_2(Pd), CH_3COOH) OH / C(=O)NH—SO_2F → (1. NaOH 2. $H^{\oplus}$) O, SO_2, NH, O

85%
m.p. 170°C

Scheme 3.17

Aryloxysulphonyl isocyanate was also used in the preparation of oxathiazinone dioxides [80] (Scheme 3.18).

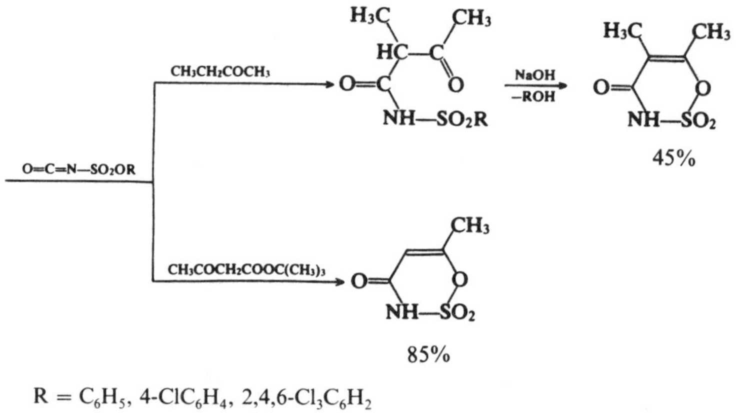

O=C=N—SO_2OR → ($CH_3CH_2COCH_3$) H_3C, CH_3, HC—C, O=C, O, NH—SO_2R → (NaOH, −ROH) H_3C, CH_3, O=, O, NH—SO_2 45%

O=C=N—SO_2OR → ($CH_3COCH_2COOC(CH_3)_3$) CH_3, O=, O, NH—SO_2 85%

R = C_6H_5, 4-ClC_6H_4, 2,4,6-$Cl_3C_6H_2$

Scheme 3.18

An alternative synthesis of 6-methyl-1,2,3-oxathiazin-4-one-2,2-dioxide in 86.5% yield was reported in 1986 [94]. It is based on the cyclocondensation of 3-oxobutanamide with sulphuryl fluoride in potassium carbonate containing aqueous acetone (6.7 : 93.3).

The method of preparation of Acesulfam K is also patented [95]; it utilizes $CH_3COCH_2CONHSO_3H$ which, when dissolved in dichloromethane and sulphur trioxide at $-40\,°C$ to $-50\,°C$ forms an adduct, that is in turn hydrolysed in methanolic KOH (pH 8–10) at $-30\,°C$.

3.3.1 Properties and reactions of oxathiazinone dioxides

Oxathiazinone dioxides are stable crystalline compounds, soluble in organic solvents and often also in water. They are monobasic acids and form neutral salts, with high melting points (Na 220 °C, K 250 °C, Ca over 310 °C). Their 5,6-dimethyl- and 5-ethyl-6-methyl derivatives can be distilled in high vacuum.

The acidic hydrogen in the molecule of oxathiazinone dioxides can be easily methylated by the ethereal solution of diazomethane to give the corresponding *N*-methyl derivative in 68% yield, as well as the *O*-methyl compound in 16 %. The use of dimethyl sulphate yields exclusively the *N*-methyl derivative (Scheme 3.19).

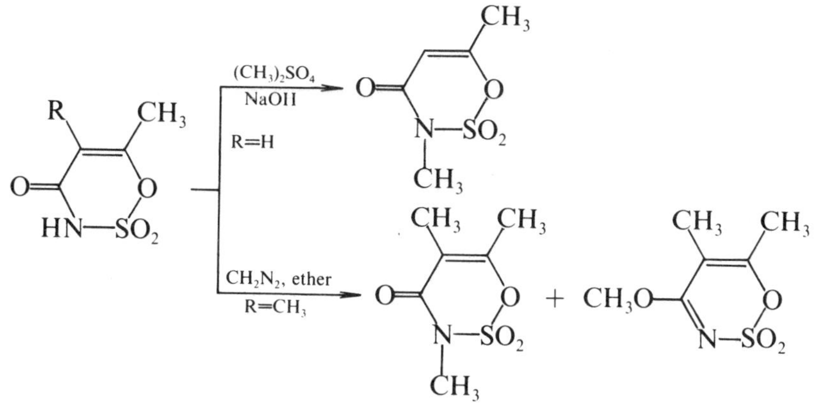

R CH3 O O HN—SO2
(CH3)2SO4 NaOH R=H
CH3 O O N—SO2 CH3
CH2N2, ether R=CH3
CH3 CH3 O O N—SO2 CH3 + CH3O CH3 CH3 O N—SO2

Scheme 3.19

The treatment of the 5,6-dimethyl derivative with phosphorus pentachloride in carbon tetrachloride affords the 4-chloro derivative, an intermediate for further reactions such as the conversion into a 4-anilino derivative (in 48 %) and subsequent bromination which results in the formation of α-bromo derivatives with an opened oxathiazinone ring (Scheme 3.20).

Extensive series of dihydrooxathiazinone dioxides have been prepared. All compounds displayed some degree of sweetness, provided that the parent skeleton of oxathiazinone was retained, as was already

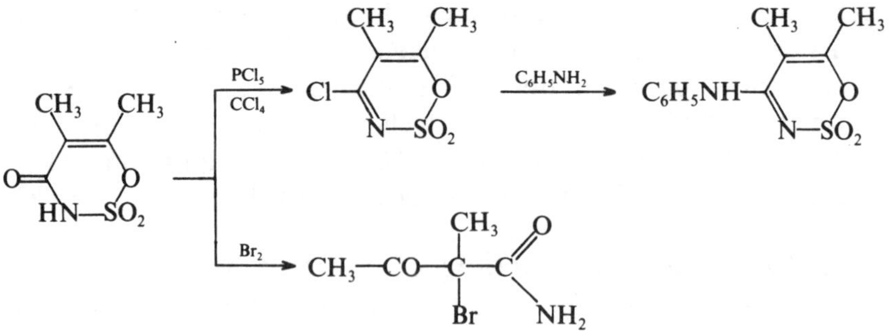

Scheme 3.20

established by Clauss and Jensen [80]. The compounds below are all insipid.

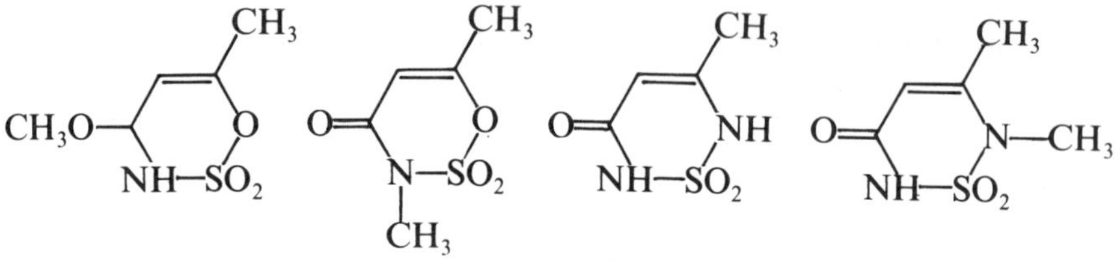

Comparative tests have shown that the potassium salt of 6-methyl-1,2,3-oxathiazin-4-one-2,2-dioxide — Acesulfam K (Table 3.3) has the cleanest and most pleasing sweet taste.

Acesulfam K is a white crystalline compound (monoclinic), the distance between the N and S atoms, which lie in the C=C plane, are 0.0125 and 0.0433 nm, respectively [19, 97]. It has no sharp melting point; when heated slowly, it begins to decompose at about 225 °C. It dissolves well in water; 1 liter dissolves as much as 270 g at 20 °C and 50 % solution can be prepared at the boiling point. The solubility is quite low in pure organic solvents, thus only 1 g will dissolve in 1 litre of absolute ethanol at room temperature. However, the solubility improves dramatically in aqueous mixtures — approximately 100 g will dissolve in 1 liter of 50 % aqueous ethanol [86]. Dilute Acesulfam K solutions are almost neutral.

The shelf life of dry Acesulfam K is exceptionally long. Samples stored for as long as 8 years looked visually perfect and their analytical data were the same as those for freshly prepared samples. Furthermore, there was no difference between samples exposed to light, and those stored in the dark. Acesulfam K is not hygroscopic [87].

The hydrolytic stability of Acesulfam K is also very good. It is stable in the pH range usually applicable for food (3–7) and a measurable degradation occurs when the pH is under 3, and at elevated temperatures. The sterilization of aqueous solutions, at pH 5, at 120 °C poses no stability problems. Acesulfam K decomposes under extreme conditions into acetone, carbon dioxide and ammonium sulphate.

The hydrothermal stability, in relation to the initial concentration of the product and the pH, have been monitored [88]. The rate of decomposition in aqueous solutions, catalysed by amino acids in model conditions, was also measured [89], as well as model binary equimolar mixtures with hydroxycarboxylic acids and dicarboxylic acids [90].

Acesulfam K, at its maximum sweetness, is 200 times sweeter than a 3 % solution of sucrose; the sweetness begins to thin out at higher than the optimal concentration. Its sweetness is delicate, with pleasant quality. There are no disturbing effects of aftertaste at low and medium (most commonly used) concentrations. Its sweetness has been rated even higher than that of sucrose. The elicited sweet sensation sets in quickly, without any induction period, and persists slightly longer than with sucrose. The relationship between the purity of the taste and the concentration of the sweetener, an important factor in the food industry, was checked in order to assess the upper concentration limit [91].

Acesulfam K has been extensively tested for possible health risks under the conditions specified by the Joint Expert Committee for Food Additives of WHO and FAO — no objections against its use have been raised. It has been certified safe for both food and drinks at ADI levels of 0 to 9 mg/kg body weight. The dose can almost completely substitute the natural sugars contained in the average diet [84]. US Food and Drug Administration legislated the use of Acesulfam K as a food additive on 28th July 1988 [92].

The measured acute toxicity of Acesulfam K is low; LD_{50} oral doses were 6.9–8 g/kg body weight in intraperitoneal administration. Acesulfam K could be added to animal feed in 10 % concentration as part of testing for subchronic toxicity; it is non-toxic, non-carcinogenic, and neither any mutagenic activity nor any teratogenic effect have so far been detected.

The human body does not metabolize Acesulfam K. The possible metabolic transformations were studied by using ^{14}C-labelled compound. Acesulfam K was excreted in the urine completely unchanged.

Apart from its sensory properties and the effect of potassium ions, Acesulfam K showed no pharmacological activity. The survey of data in Table 3.1 exemplifies the applications of Acesulfam K in the food, drinks and as an alternative 'tabletop' sweetener. The last mentioned preparations encompass all forms of domestic sugar substitutes used in food preparation, such as tablets (also effervescent) powder, sachet packets, granular form and solutions.

Tablets can be easily prepared when Acesulfam K combined with a suitable diluent, fòr example by mobile forms of carboxymethyl cellulose. Such pellets are very compact, have good mechanical properties and still dissolve quite fast because of the excellent solubility of Acesulfam K. Effervescent tablets are best prepared by adding $NaHCO_3$ in combination with tartaric acid. Such tablets are stable provided, they are stored in dry conditions.

The insoluble formulation of Acesulfam K, designed as additive for pastries, tarts, fruit salads and similar products, contains microcrystalline cellulose; such a mixture does successfully imitate the sugar powder based dessert toppings.

Good water solubility allows the preparation of stock solutions which, when stabilized by suitable buffers (in the range pH 5.5–6), are stable for years [87].

The most important single application of Acesulfam K is in the preparation of non-alcoholic beverages to which it is added either as a powdered beverage mix or as a liquid sweetener. It is equally applicable for use in beer, wine and hard drinks. Another popular application is in fruit-based products where it blends well with hydrocolloidal diluents.

When applied in pastries, it must be combined with saccharides in order to create the typical aroma, form and flavour. A simple combination of Acesulfam K with sorbitol generates a typical flavour that is preferred by sweet lovers [86]. Acesulfam K, in combination with diluents, can be added to chewing gum as well as to a wide variety of confectioneries.

The characteristic properties of Acesulfam K can be exploited in the manufacture of other products as well, as is exemplified by mouth cosmetic preparations in which its compatibility with most aromatic and flavour constituents has been utilized. Toothpastes and mouth lotions are amongst the typical applications. Its sweet taste is not made prominent, instead it is used for masking the unpleasant taste of some constituents. Its solubility in aqueous alcohol is just sufficient to reach the required sweetness [86]. Acesulfam K performs a similar masking function in pharmaceutical applications, specially in drugs for diabetics.

Although no serious health hazards seem to be associated with the use of synthetic sweeteners, fast and reliable analytical methods are essential, both for the quantitative and qualitative control of products and for the purity control of the additive.

By analogy with saccharin and cyclamates, Acesulfam K is detected by thin-layer chromatography on polyamides — the liquid phase used is a 5 : 5 : 1 mixture of xylene, *n*-propanol and formic acid [75]. Zache and Gründing [47] separated acesulfam, saccharin and aspartame by liquid chromatography and capillary isotachophoresis. The method is also suitable for repeated routine analysis as well as for purity control of non-alcoholic beverages [93].

Fast and precise HPLC chromatography could be finely adjusted for the analytical determination of Acesulfam K in drinks, food and tooth paste [96]. High-performance ion chromatography was used for similar purpose by Biemer [78].

3.4 ASPARTAME AND ITS ANALOGUES

The discovery of the sweet taste of aspartame (APM) was just as much accidental [102–104] as the discoveries of saccharin and cyclamates. In December 1965 J. Schlatter, working in the research laboratories of Searle (Skokie, Illinois), was recrystallizing Asp-Phe-OMe, an intermediate of the *C*-terminal tetrapeptide Trp-Me-Asp-Phe-NH_2 of the hormone gastrin. The methanolic solution of the aspartame bumped whilst being heated and some of it splashed on the outside of the flask — a small amount of the powder got on his fingers. Some time later, Schlatter licked his finger in order to pick up a piece of paper — he noticed a very strong sweet taste. Although Schlatter and Mazur did not know it at the time, essentially the same synthesis had been carried out in the pharmaceutical laboratories of ICI [105], where the sweet taste of the compound was not noticed. Since the discovery at G. D. Searle and Co., who were also the first to market aspartame, numerous papers and research reports were published. In Czechoslovakia, aspartame hydrochloride has been marketed under the brand name Usal. The Usal technology originated in the laboratories of the Czechoslovak Academy of Sciences and at the pharmaceutical company Léčivá.

In the absence of any structure–sweetness relationship with predictive power outside a specific family of sweeteners, the studies of the synthesis and testing of structural analogues of aspartame commenced immediately. The results made a significant contribution to the theory of sweetness. The basic problem itself remains largely unsolved as there is an enormous structural variation at the receptor site with which the sweeteners interact [106].

eet dipeptides, together with other low-calorie natural and syn-tic sweeteners are suitable for diabetics and adipose persons [107–110]. The acceptance of the new sweetener by the food and drinks industry was exceptionally good. It is marketed under the brand name Nutra Sweet — this high potency, low-calorie ingredient has created considerable interest within the industry and with consumers alike in a very short period of time.

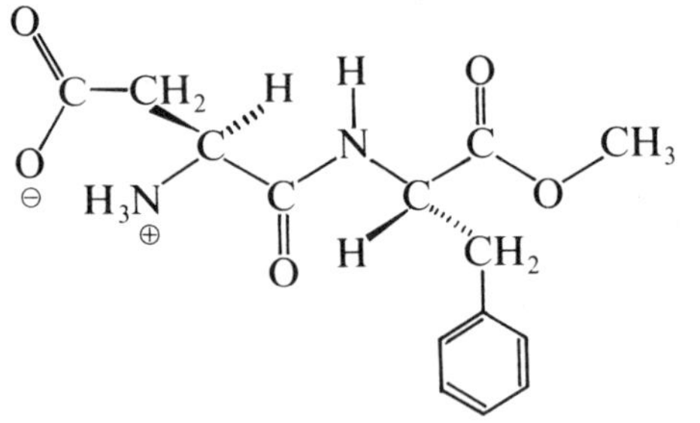

L-asparyl-L-phenylalanine methyl ester — aspartame

Ideally, a high-intensity sweetener should meet the following standards: be safe, have a clean, sweet taste, be low in calories, have a high sweetening potency, possess a reasonable stability, should not promote tooth decay, be realistically priced, be easily metabolized, be non-laxative and readily available. Aspartame is a prime choice amongst the low-calorie sweeteners because it meets the requirements almost ideally.

Since aspartame has some very specific characteristics, the food technologists should be fully aware of its advantages and its limitations [111].

3.4.1 Synthesis of aspartame and its analogues

There are numerous papers that deal with the synthesis of aspartame. In general, the synthetic procedures that lead to aspartame can be broken down to two groups, one employing the selective removal of the protecting groups of the β-carboxy of the aspartic acid, the other utilizing cyclic derivatives, such as anhydrides or thioanhydrides. The original synthesis of aspartame, carried out by the discoverers, involved a condensation between the β-benzylester of the *N*-benzyloxycarbonyl aspartic acid and methyl ester of phenylalanine *via* either the active ester or a mixed anhydride. The protecting groups were removed by a catalytic reduction (Scheme 3.21).

The synthetic pathway was chosen in order to prevent the formation of the β-isomer and thus simplify the purification of the final product.

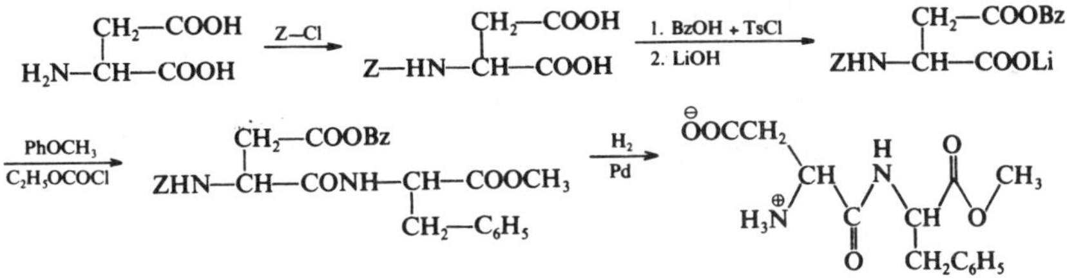

$Z = C_6H_5CH_2{-}O{-}CO$
$Bz = C_6H_5CH_2$

Scheme 3.21

However, in spite of this, the method is obviously too costly for a large scale production. If the availability and the price of the starting material is made a factor, the most attractive approach to aspartame appears to be one that is based on the ammonolysis of the internal aspartic anhydride by the protected amino group [113–119] (Scheme 3.22); utilizing the standard protecting groups used in peptide synthesis. The anhydride is utilized in the form of various salts [113–117], the ammonolysis itself is carried out in pure organic solvents, in aqueous solvents or even in water alone at controlled pH [116] (Scheme 3.22).

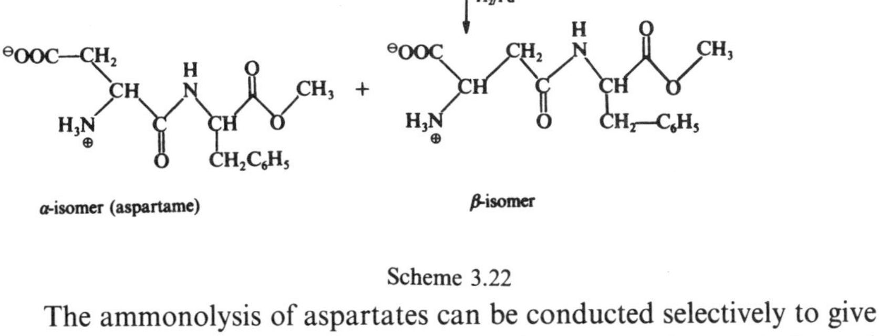

Scheme 3.22

The ammonolysis of aspartates can be conducted selectively to give the α-isomer at lower temperatures. Similarly, the α-isomer is formed when an excess of the amino component is used, or alternatively, when the reaction mixture contains various additives [120]. These can be any of the following:

(a) acetic acid [120], carbon dioxide [121, 122], *o*-phosphoric acid [123]
(b) mixtures of weak acids and lower aliphatic alcohols [121]
(c) mixtures of strong acids and alcohols [124]
(d) solutions of inorganic bases [125]

The described synthetic methods are, of course, far from perfect. Although simplicity is their strong point, the necessity to purify the main product in order to remove the bitter β-isomer and other very similar by-products, is a decisive drawback [126, 127].

The strategies devised for the separation of the α- and β-isomers of aspartame employ several approaches.

The dissimilar solubilities of their respective salts with strong inorganic [115, 128] or aromatic acids [129] lend themselves as a method of separation; salts of the α-isomer being usually less soluble in water or aqueous organic solvents than the corresponding salts of the β-isomer.

Another separation strategy capitalizes on the decreased reactivity of the β-isomer with ketones. When reacted with acetone, the α-isomer of aspartame forms easily the corresponding acetone-soluble 4-imidazolinone (Scheme 3.23), whilst the β-isomer fails to react under the same conditions and precipitates from the solution [130, 131]. The problem in this approach is the necessity to hydrolyse the rather stable 4-imidazolinone at elevated temperatures, when a partial thermal decomposition of the liberated labile aspartame cannot be avoided [132].

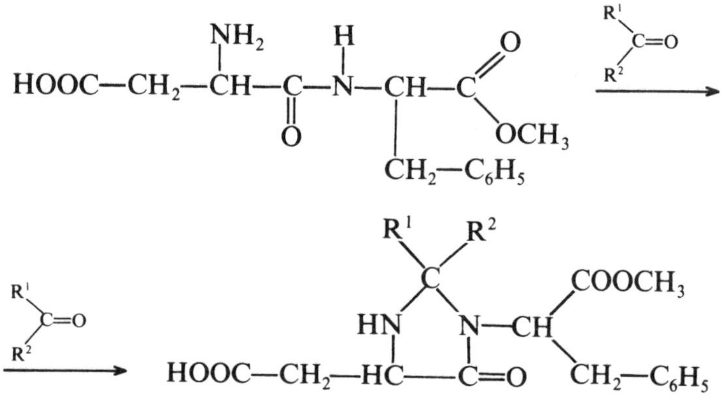

$$HOOC{-}CH_2{-}CH(NH_2){-}C({=}O){-}NH{-}CH(CH_2{-}C_6H_5){-}C({=}O)OCH_3 \xrightarrow{R^1R^2C{=}O}$$

$$\xrightarrow{R^1R^2C{=}O} \text{4-imidazolinone: } C(R^1)(R^2) \text{ ring with } HN,\ N{-}CH(COOCH_3)(CH_2{-}C_6H_5),\ HOOC{-}CH_2{-}HC{-}C{=}O$$

$R^1 = CH_3$; $R^2 = CH_3, C_2H_5$; $R^1, R^2 = (CH_2)_5$

Scheme 3.23

An ingenious large-scale separation method [133] is based on the reaction of the *N*-protected or unprotected aspartic acid with the ester of 2-isocyanato-3-R-propionic acid (R = phenyl, cyclohexyl). However, the method uses phosgene at elevated temperatures and thus imposes stringent safety requirements on the apparatus. Furthermore, the final product is contaminated with the β-isomer (Scheme 3.24).

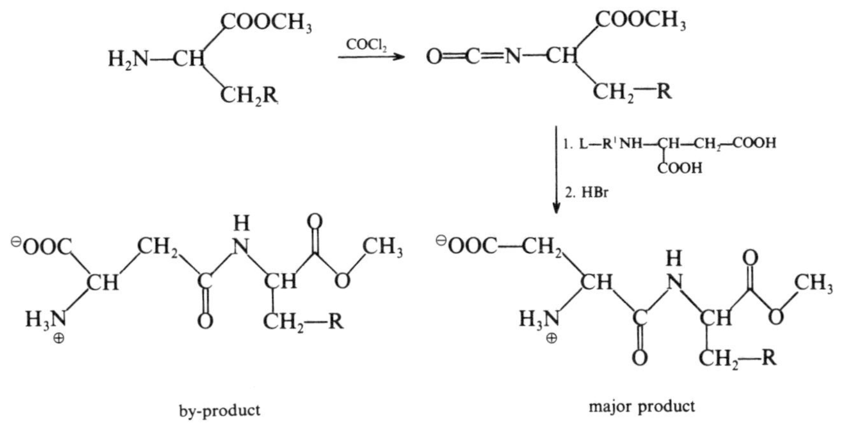
COOCH$_3$
H$_2$N—CH
CH$_2$R
COCl$_2$
O=C=N—CH
COOCH$_3$
CH$_2$—R
1. L—R^{1}NH—CH—CH$_2$—COOH
COOH
2. HBr
by-product
major product

R = phenyl, cyclohexyl

R^1 = H, $C_6H_5CH_2$—O—CO

Scheme 3.24

An aspartame preparation which utilizes the reaction of the *N*-protected aspartic acid with formaldehyde has been described [134–136]. Oxazolidinines formed in the first step condense with L-phenylalanine in the next step; the hydrolysis of the product leads to aspartame (Scheme 3.25).

α-Aspartyl derivatives are accessible from the *N*-carboxyanhydride of the β-*O*-protected derivatives of aspartic acid [137–139]. Problems arise during the ammonolysis of the anhydride conducted at basic pH, the ester bond is also prone to hydrolyse under such drastic conditions. Japanese authors [140] synthesized aspartame starting with maleic acid dichloride and obtained the product in the D,L-L-form.

Pietsch [141] in his synthesis of aspartame followed a different strategy. Both the amino group and the β-carboxylic group of aspartic acid protects a lactam ring which is created by the oxidation of *S*-4-vinyl-2-azetidinone. The acid then undergoes, in its protected form, a condensation with the methyl ester of phenylalanine catalysed by dicyclohexylcarbodiimide (DDC). The protected lactam ring is then opened by the action of hydrochloric acid (Scheme 3.26).

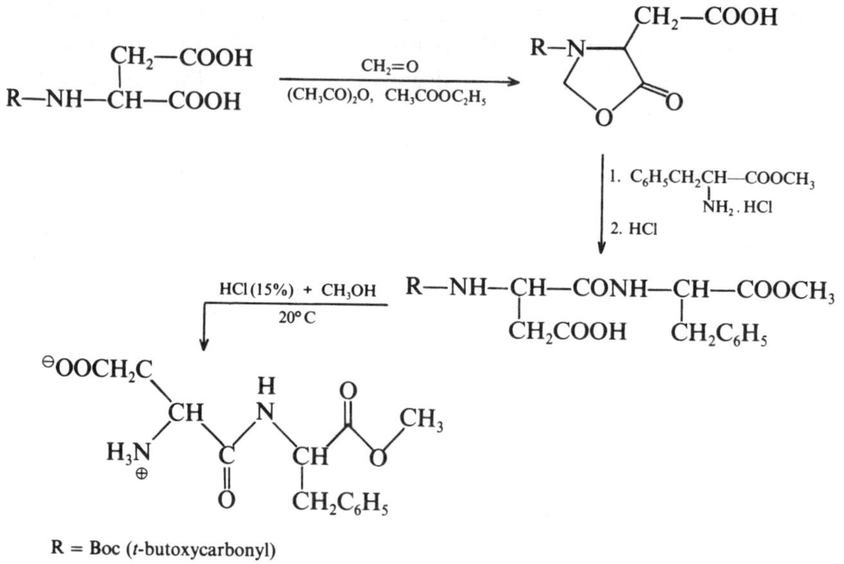

$R{-}NH{-}CH(CH_2{-}COOH){-}COOH$ → ($CH_2{=}O$; $(CH_3CO)_2O$, $CH_3COOC_2H_5$) → 3-R-5-oxo-oxazolidine-4-acetic acid ($CH_2{-}COOH$)

↓ 1. $C_6H_5CH_2CH(NH_2 . HCl){-}COOCH_3$ 2. HCl

$R{-}NH{-}CH(CH_2COOH){-}CONH{-}CH(CH_2C_6H_5){-}COOCH_3$

↓ HCl(15%) + CH_3OH, 20° C

$^{\ominus}OOCH_2C{-}CH(H_3\overset{\oplus}{N}){-}C(=O){-}NH{-}CH(CH_2C_6H_5){-}C(=O){-}O{-}CH_3$

R = Boc (*t*-butoxycarbonyl)

Scheme 3.25

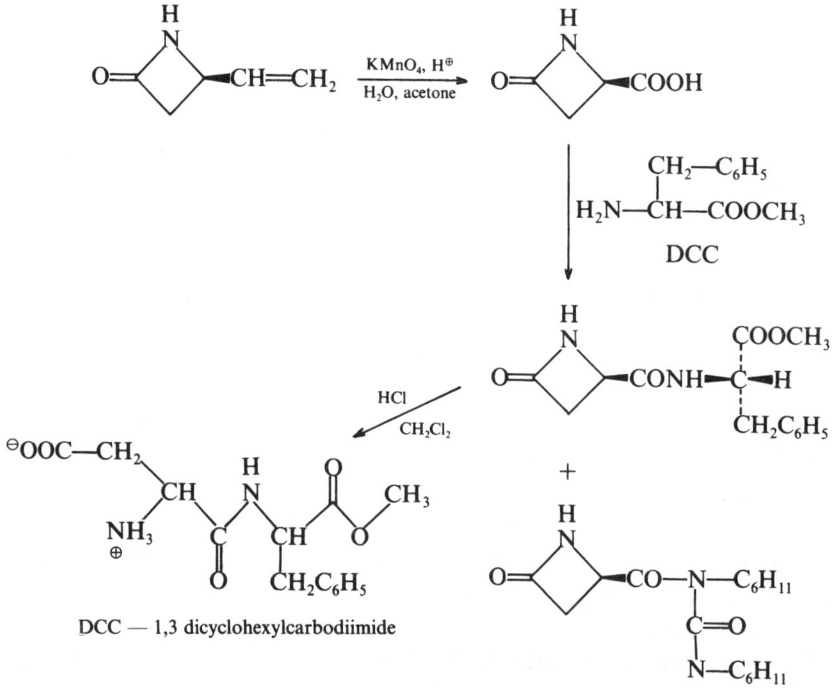

4-vinyl-azetidin-2-one ($CH{=}CH_2$) → ($KMnO_4$, $H^{\oplus}$; H_2O, acetone) → 2-oxo-azetidine-4-carboxylic acid ($COOH$)

↓ $H_2N{-}CH(CH_2{-}C_6H_5){-}COOCH_3$, DCC

2-oxo-azetidin-4-yl$-CONH{-}C(H)(COOCH_3)(CH_2C_6H_5)$

+

2-oxo-azetidin-4-yl$-CO{-}N(C_6H_{11}){-}C{=}O{-}N{-}C_6H_{11}$

↓ HCl, CH_2Cl_2

$^{\ominus}OOC{-}CH_2{-}CH(\overset{\oplus}{N}H_3){-}C(=O){-}NH{-}CH(CH_2C_6H_5){-}C(=O){-}O{-}CH_3$

DCC — 1,3 dicyclohexylcarbodiimide

Scheme 3.26

High yield of aspartame was obtained in a single-step procedure starting with the anhydride of *N*-thiocarboxyaspartic acid (Scheme 3.27). Moreover, the regiochemistry of the reaction remained fully under control [142].

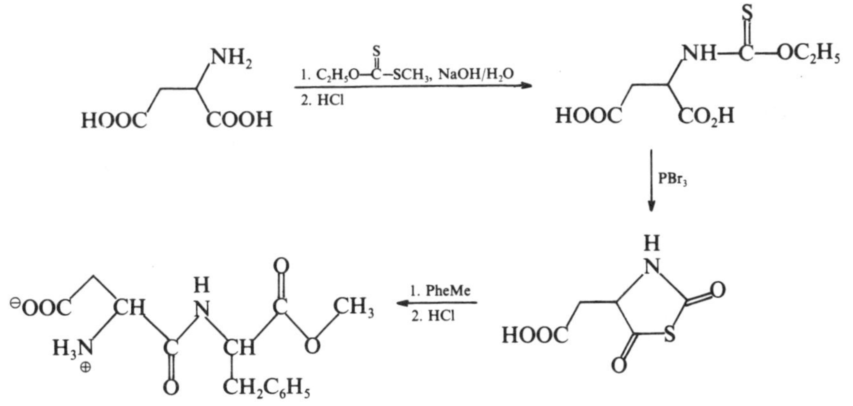

Scheme 3.27

The conventional synthetic methods, developed from the chemistry of peptides, have been also employed in attempts to prepare aspartame and its analogues. One such scheme, the carbodiimide method, utilizes either mixed anhydrides or activated esters [143–145]. In a variation of this method, the condensation was catalysed by trifluoroacetoxysuccinimide [146], which facilitated a quick and complete conversion of the *N*-β-*O*-disubstituted carboxy component to the corresponding α-*N*-oxysuccinimide ester. The ester was ammonolysed in the final step.

Several patents [147–153] deal with the modified procedures for the removal of the protecting groups in the course of the synthesis of free, sweet dipeptide esters. The utilization of enzymes in the synthesis of aspartame and its analogues was claimed [149, 150, 154] for coupling of the β-benzyl ester of *N*-benzyloxycarbonylaspartic acid with phenylalanine methyl ester (enzyme thermolysine). The method allowed the utilization of cheap racemic phenylalanine ester [102, 155, 156]. Thermolysine was also employed in its immobilized form [157] in a range of organic solvents; the best results were achieved in ethyl acetate [158].

Scheme 3.28 outlines the synthesis of a series of sweet esters of the L-aspartyl-1-aminocyclopropanecarboxylic acid [159].

In the first step, 1-aminocyclopropanecarboxylic acid was converted to its esters *via* the acid chloride. Next, the esters were coupled through

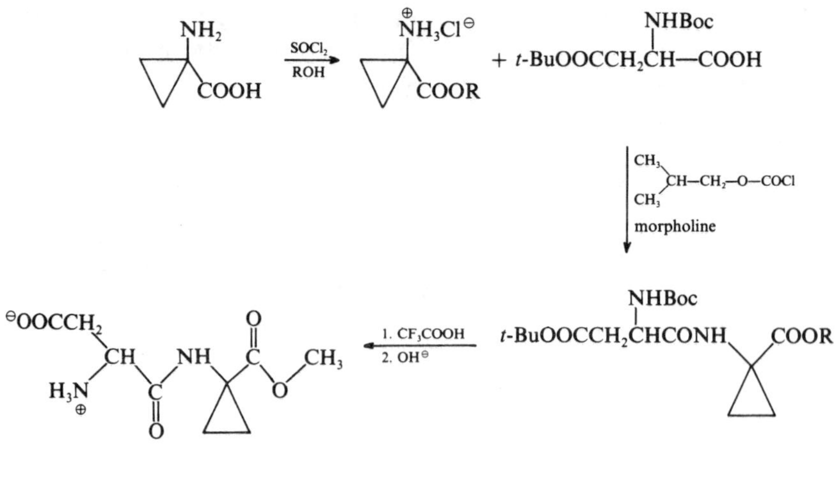

Scheme 3.28

$R = CH_3, C_2H_5, n\text{-}C_3H_7, i\text{-}C_3H_7, n\text{-}C_4H_9, i\text{-}C_4H_9$

the amino group with the protected L-aspartic acid. Subsequent removal of the protecting groups furnished esters of L-aspartyl-1-aminocyclopropanecarboxylic acid. The structure of the sweetest ester, *n*-propyl, which is 300 times sweeter than sucrose, was confirmed by an X-ray analysis [159]. In order to assess the role of chirality on the degree of sweetness at the carboxylic terminus, a series of didehydro analogues [160] have been prepared (Scheme 3.29).

The most difficult step in concatenation of reactions leading to didehydroaspartame was the methanolysis of the unsaturated azlactone. It proceeded smoothly and unequivocally only when catalysed by an equivalent of 4-dimethylaminopyridine.

3.4.2 Structure–taste relationship of aspartame

The sweet taste of aspartame could not have been predicted from the sensory properties of the constituent amino acids; L-aspartic acid is almost tasteless or slightly acidic, whilst L-phenylalanine is bitter [161]. Of all the possible diastereoisomers, only the one with both acids in the L-form is sweet; the D,D-enantiomer, D,L- and L,D-diastereoisomers have slight bitter taste [103]. The situation clearly demonstrates the high stereospecificity in the interaction with the sweet receptor. In addition, the methyl ester structural element proved essential for the sweetness; L-α-aspartyl-L-phenylalanine is tasteless (Table 3.4).

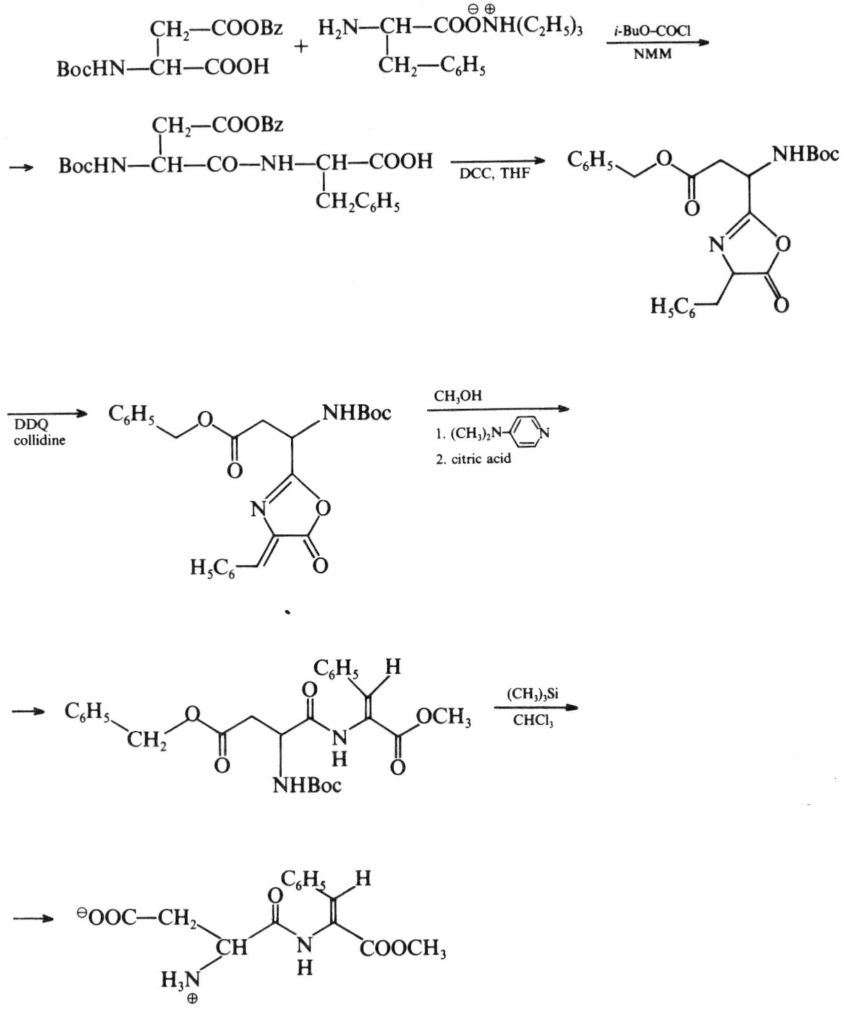

$$BocHN{-}CH(CH_2{-}COOBz){-}COOH + H_2N{-}CH(CH_2{-}C_6H_5){-}COO^{\ominus}\overset{\oplus}{N}H(C_2H_5)_3 \xrightarrow[NMM]{i\text{-}BuO\text{-}COCl}$$

$$\rightarrow BocHN{-}CH(CH_2{-}COOBz){-}CO{-}NH{-}CH(CH_2C_6H_5){-}COOH \xrightarrow[DCC,\ THF]{}$$

$$\xrightarrow[DDQ,\ collidine]{} \xrightarrow[1.\ (CH_3)_2N\text{-}C_5H_4N;\ 2.\ citric\ acid]{CH_3OH}$$

$$\rightarrow \xrightarrow[CHCl_3]{(CH_3)_3Si}$$

$$\rightarrow {}^{\ominus}OOC{-}CH_2{-}CH(\overset{\oplus}{N}H_3){-}CO{-}NH{-}C(COOCH_3){=}CH{-}C_6H_5$$

NMM — *N*-methylmorpholine

DDQ — 2,3-dichloro-5,6-dicyano-1,4-benzoquinone

Scheme 3.29

Table 3.4 — Relative sweetness of aspartame, its derivatives and analogues [103]

Compound	Relative sweetness[a]
α-(L)-Asp-(L)-Phe-OMe	180
α-(L)-Asp-(L)-Met-OMe	100
α-(L)-Asp-(L)-Tyr-OMe	10
α-(L)-Asp-(L)-Phe-OEt[b]	10
α-(L)-Asp-(L)-Phe-OPr-*n*[b]	1
α-(L)-Asp-(L)-Phe-OPr-*i*[b]	1
α-(L)-Asp-(L)-Phe-OBu-*t*[b]	1
α-(L)-Asp-(L)-Phe-NH_2	0
α-(L)-Asp-(L)-Phe	0
β-(L)-Asp-(L)-Phe-OMe	0
α-(L)-Asp-(L)-Phe-OMe	0
α-(L)-Glu-(L)-Phe-OMe	0
D,L-Ama-(L)-Phe-OMe[c]	230
α-(L)-Asp-(D)-Phe-OMe	0
α-(D)-Asp-(D)-Phe-OMe	0
α-(D)-Asp-(L)-Phe-OMe	0

[a] The sweetness of sucrose = 1.
[b] Ref. [113, 143].
[c] Ref. [162, 163].

Mazur's group synthesized more than 50 aspartame analogues and was the first to investigate the structure–sweetness relationship. They found that in order for a dipeptide to be sweet, its molecule should possess the following structural characteristics:

(a) free amino and carboxylic groups separated by a certain distance
(b) the required configuration at the asymmetric carbon atom and
(c) an ester group at the *C*-terminal of the dipeptide

These requirements were believed to be essential for the manifestation of the sweet taste (see Table 3.4). However, it has been shown later that even within the aspartyl class of sweeteners, this was not universally true.

Interesting studies were carried out in which the aspartyl moiety was replaced by structurally related amino acids. Thus the lower homologue — aminomalonylphenylalanine methyl ester [162] was sweet (230 times sweeter than sucrose), but the higher homologue — glutamylphenylalanine methyl ester [163] was neutral.

In the course of other structural studies on aspartyl dipeptide esters, the structure of aspartame itself elicited the most powerful sweet response; both the lower homologue of L-phenylalanine — methyl ester of aspartyl methylglycine (140 times sweeter than sucrose) [164] and the higher homologue — methyl ester of aspartyl homophenylalanine (100 times sweeter than sucrose) [104] were inferior sweeteners.

It has been shown later that the L-Phe-OMe moiety can be successfully substituted by alkyl- and arylamines [165], by esters of D- and L-amino acids [103; 166, 167], by derivatives of D-amino acids [168, 169] as well as by malonic diesters [170]. However, any structural modifications that involved the peptide bond, for example methylation, resulted in the disappearance of the sweet taste [171].

The quantitative assessment of the sweet potency of L-α-aspartyl dipeptides has been undertaken. The study evaluated the structural electronic and hydrophobic parameters [172, 173]. The conformations of L-α-Asp-L-Phe-OMe likely to be responsible for the sweetness were studied by the combination of NMR methods and potential energy [174] and by the synthesis and the taste monitoring of L-Asp-Δ^Z-Phe-OMe (dehydroaspartame) [160].

In a study aimed at relating the sweet properties to the molecular structural features and thus creating a model with predictive powers, Ariyoshi [175] proposed that the correlation of L-aspartyl dipeptide analogues depended on the common features that could be easily visualized by using the Fischer projection formula of dipeptides

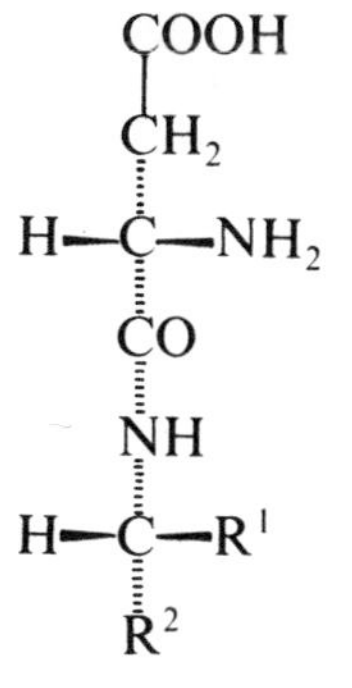

COOH–CH_2–C(H)(NH_2)–CO–NH–C(H)(R^1)–R^2

sweet

COOH–CH_2–C(H)(NH_2)–CO–NH–C(R^1)(H)–R^2

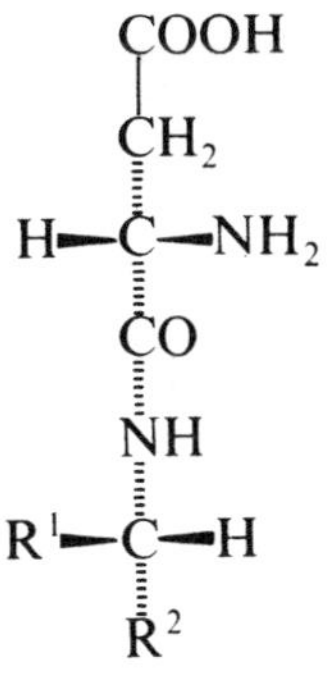

not sweet

R^1 = small hydrophobic group
R^2 = larger hydrophobic group

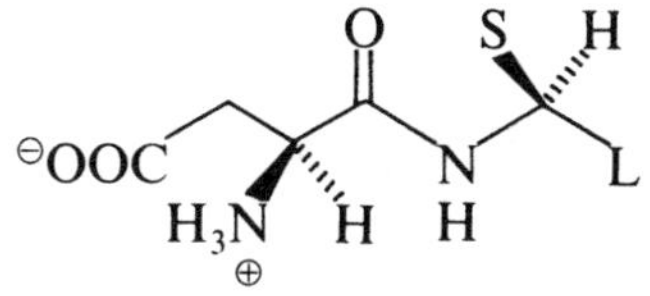

not sweet

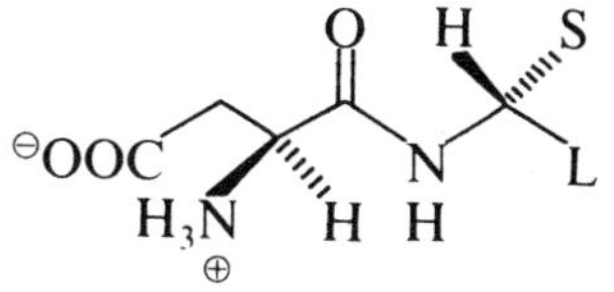

sweet

According to Ariyosi [176], a small group (S) in aspartyl dipeptides participates in the formation of hydrophobic interactions with the corresponding partner in the sweetness receptor, and the larger (L) hydrophobic group favours certain desired conformations and keeps in balance the overall hydrophilicity and hydrophobicity of the molecule.

Another factor that has been recognized concerns the space-filling properties of the groups attached to the carbon atom that carries the amidic nitrogen atom. It appears that sweet aspartyl derivatives assume the spatial form that binds to the active centre inside a deep furrow [174], a pocket [176] or a hydrophobic pipe [177] on the receptor site.

Specially synthesized aspartyl tripeptide esters [178] helped to confirm the assumed hypothetical deep pocket shape of the receptor site [176].

The sweetness requirements in the tripeptide series dictate that the second amino acid be in the D-configuration, and that the R^2 group be a small, compact alkyl group (methyl, ethyl, 2-propyl).

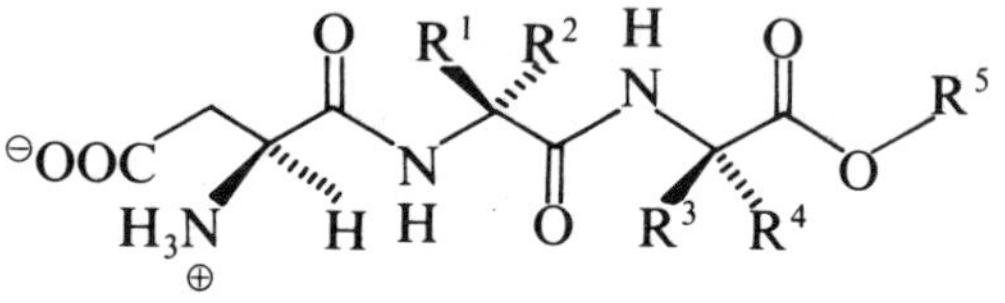

Moreover, the third amino acid must have the L-configuration. The small R^2 group participates in the hydrophobic interactions with the receptor site and enhances the sweetness.

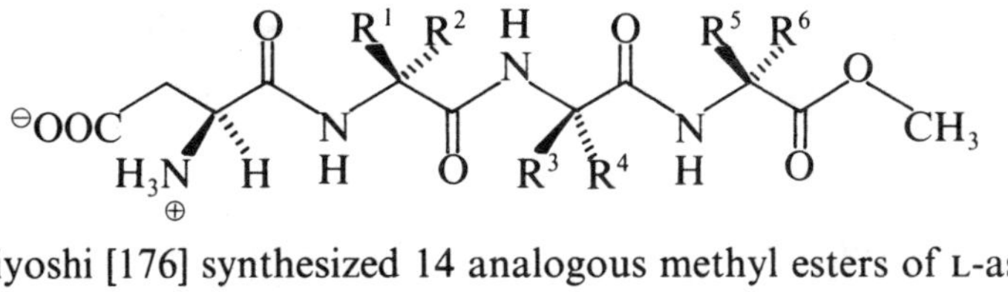

Ariyoshi [176] synthesized 14 analogous methyl esters of L-aspartyl tetrapeptides and established that similar structural considerations, applicable to the tripeptides, also apply in this series [178].

Ariyoshi pursued the structure–sweetness studies even further in the pentapeptide series where suddenly the continuity terminated — none of the relationships derived for di-, tri- and tetrapeptide series were applicable. The reason is that the pentapeptides are probably too bulky to fit into the pocket-shaped receptor site.

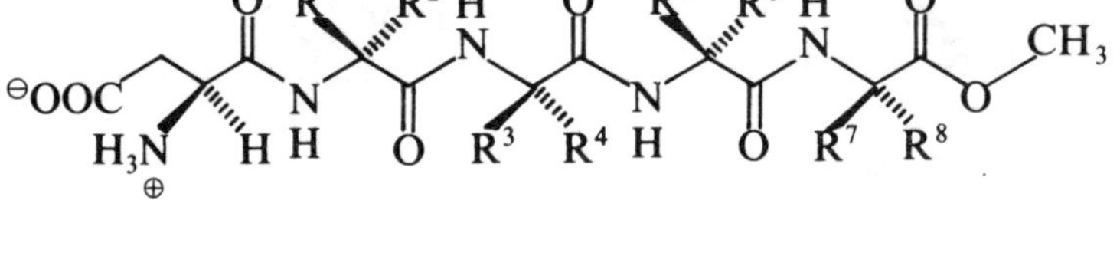

The list of sweet dipeptides expanded once more when the sweet taste of other esters, such as L-α-aspartyl-L-threonine, L-α-aspartyl-L-methionine and L-α-aspartyl-*S*-alkyl-L-homocystine [143, 181] was discovered. ^1H NMR studies helped to establish the best conformation in H-Asp-Met-OMe [181]. The introduction of bulky substituents in position 4 of the phenyl group of the aspartame drastically reduces, or even eliminates the sweet taste. The effect of similar substitution in position 2 is much less pronounced (Table 3.5). On the other hand, partial or complete hydrogenation of the aromatic ring affects only the taste intensity [133, 166, 184–187].

Table 3.5 — Relative sweetness of aspartame derivatives with substituted phenyl group

Configuration at the terminus		Position X			Relative
N	C	2	3	4	sweetness[a]
L	L	H	H	OH	10–40[b]
L	D	H	H	OH	0
L	D,L	H	H	OH	10–40[b]
L	L	H	H	OCH_3	1–5[b]
L	L	H	OH	H	10–40
L	D	H	OH	H	0
L	L	H	OCH_3	H	10–40
L	D,L	H	OCH_3	H	10–40
L	L	H	H	OH	10–40
L	L	OH	H	H	10–40
L	D	OH	H	H	0
L	D,L	OH	H	H	10–40
L	D,L	OCH_3	H	H	80–200
L	L	H	OH	OH	1–5
L	L	H	OH	OCH_3	0
L	L	H	OCH_3	OH	1–5
L	L	H	OCH_3	OCH_3	0

[a] Ref. [182].
[b] Ref. [103].

Whilst correlating the intensity of sweetness with various structural elements present in the molecule, Mazur and his group [166] discovered that the phenyl group in the *C*-terminal amino acid can be replaced by an alkyl group. It has been also demonstrated [188] that the intensity of sweetness of aspartyl dipeptides varied with the size, shape and hydrophobicity of the R-residue (Table 3.6). Accordingly, methyl esters of the L,L-form of such dipeptides were not sweet until the number of carbon atoms in R-chain reached 4.

The situation changes when the second amino acid has a D-con-

Table 3.6 — Esters of α-aspartyl dipeptides[a]

$$H_2N—CH(CH_2—COOH)—CONH—CH(R)—COOCH_3$$

R	Relative sweetness	References	R	Relative sweetness	References
Methyl	0	[166]	*n*-Amyl	50	[166, 168]
i-Propyl	0	[166]	*i*-Amyl	60	[166, 168]
n-Butyl	40	[166, 168]	*n*-Hexyl	75	[166, 168]
i-Butyl	0	[166]			

[a] L,L — configurations.

figuration. Now aspartyl dipeptides are sweet only when the R-group is kept small and the ester group is bulky [189].

Two configurational analogues of the methyl ester of L-α-aspartyl-L-phenylalanine, namely the L-aspartyl-D-alanine benzyl ester and L-aspartyl-D-α-aminobutyric acid benzyl ester (Table 3.7) are sweet, as apparently their two hydrophobic groups take up the spatial orientation similar to that of the hydrophobic groups of aspartame.

Table 3.7 — Taste of esters of L-aspartyl-D-α-alkylglycine and L-aspartyl-α,α-dialkylglycine

$$^{\ominus}OOC—H_2C—CH(H_3N^{\oplus})—C(=O)—NH—C(R^1)(R^2)—C(=O)—O—R^3$$

R^1	R^2	R^3	Taste
CH_3	H	$CH_2C_6H_5$	sweet
C_2H_5	H	$CH_2C_6H_5$	sweet
CH_3	CH_3	$CH_2C_6H_5$	tasteless
CH_3	CH_3	CH_3	sweet
C_2H_5	CH_2CH_3	CH_3	tasteless
H	H	H	tasteless

The results of all studies on the structure–sweetness relationship indicate that the influence of conformation on the sweetness is more profound than that of the chirality [189]. The authors also investigated the homologous series of methyl esters of L-aspartyl-α-aminocycloalkanecarboxylic acids and concluded that there was a definite relationship between the effective volume of the side-chain and the elicited sensory properties.

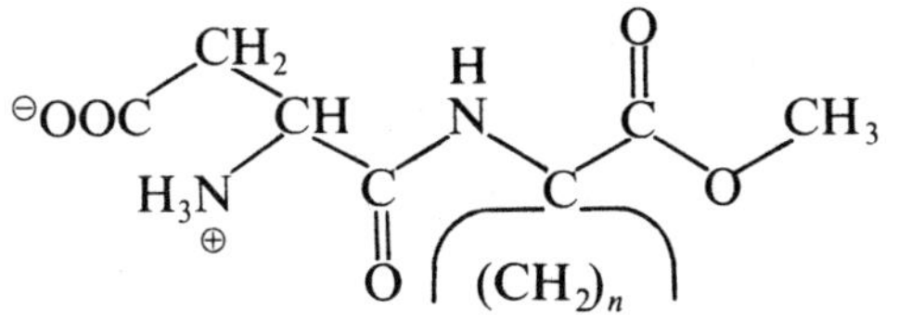

n = 2, 3, 4 — sweet, n = 5, 6 — bitter, n = 7 — tasteless

The authors [190] proposed that the interaction of aspartame and its analogues could be rationalized in terms of a three-point model. The active sites of the sweet dipeptide molecule were represented by a zwitterionic polar segment, by a hydrophobic part with specific dimensions and surface, and by the peptide bond which is capable of forming hydrogen bonds.

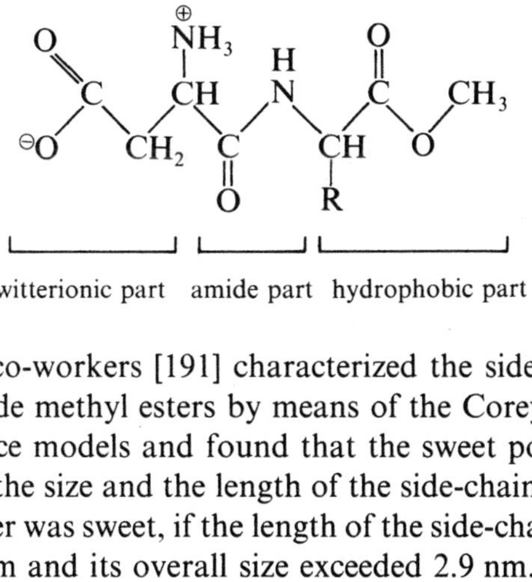

zwitterionic part amide part hydrophobic part

Brussel and co-workers [191] characterized the side-chain of 28 L-aspartyl dipeptide methyl esters by means of the Corey–Pauling–Koltum atomic space models and found that the sweet potency could be correlated with the size and the length of the side-chain. Thus a dipeptide aspartyl ester was sweet, if the length of the side-chain was between 0.48 and 0.88 nm and its overall size exceeded 2.9 nm.

Van de Heijden's group [172] used STERIMOL constants (L, B_1–B_4) to characterize the side-chain R and correlated them with the sweetness potency. They found that dipeptides were sweet when L was within limits $0.5 < L < 0.62$ nm. If the parameter L exceeded these limits, the sweetness was retained only if at the same time the parameter B_4 was within certain ranges, i.e. $B_4 > 0.45$ nm ($L < 0.5$ nm), or $B_4 < 0.72$ nm ($L > 0.62$ nm). The sweetening power, which is defined as the logarithm of sweetness, correlates significantly with both parameters (L, B_4). Furthermore, the authors established that the third binding site δ was at the centre of the phenyl group and measured the relative distances: AH—$\delta = 0.45$ nm, B—$\delta = 0.73$ nm.

Lelj and co-workers [174] consider that aspartame exists in a combination of two rotameric series D_{I}, D_{II}, D_{III} and F_{I}, F_{II}, F_{III}, and only one of these — $F_{I}D_{II}$ — interacts with the receptor. A *trans*-orientation of the ammonium $\overset{\oplus}{N}H_3$ and carboxy $COO^{\ominus}$ groups disqualifies the three rotamers with D_{I}. Further analysis was concentrated on the combinations $F_{I}D_{II}$ and $F_{II}D_{II}$; the former was selected as it had similar spatial properties as those of sweet D-leucine. The reason for this selection was based on the assumption that both D-leucine and aspartame interact with the same active site at the receptor, described as a narrow split between two sub-units of the receptor, one being responsible for the fixation of the sweet molecule and the other for the generation of the nerve impulse. The disappearance of the sweetness in *N*-alkylated amides was explained in terms of conformational analysis. The postulates were also based on the interpretation of NMR spectra and calculations of potential energy. Van der Heijden and co-workers disputed Temussi's conclusions [192], and suggested that the $F_{II}D_{II}$ conformation is more likely to interact with the receptor. He pointed out other corroborative evidence in the literature [103, 166, 191]. The ester group and the side-chain R serve as the locking groups in the $F_{II}D_{II}$ conformation, and thus their geometrical parameters affect the probability and character of the interaction and hence the relative sweetness.

Van der Heijden and co-workers [193] employed Temussi's parameters in his calculations and were able to show that the positional angles α, δ and ω vary between 125–155 °C with respect to the AH–B part of the molecule, due to the flexibility of large substituents, and that S must not exceed 0.25 nm. Low value of the parameter S limits also the maximum height of the receptor to 0.25 nm.

Recently the SIMCA method, one of the pattern recognition techniques, has been applied to the structure–taste studies on the L-aspartyl dipeptides [194]. An attempt has been made to classify the sweet and bitter class dipeptides by the SIMCA pattern recognition method using the physico-chemical descriptors such as molar refractivity and four kinds of STERIMOL parameters. The method suggested that two factors, shape and size, of the *C*-terminal amino acid moiety R in dipeptides were extremely important for modelling the overall taste qualities.

Shinoda and Okai [195] proved successfully that a trifunctional unit was responsible for the sweetness in aspartame type molecules — it consists of an electronegative sub-unit (B), an electropositive sub-unit (AH) and a hydrophobic part (X). The authors established that the AH–X component elicited bitter taste, but the taste changed to sweet when the sub-unit (B) took part in the interaction with the receptor.

In an attempt to prepare hydrolytically stable α-aspartyl type sweetener, the Japanese authors [169] synthesized a novel type of sweet molecules. In order to prevent the formation of substituted dioxopiperazines, they introduced into the chain several randomly placed methylene groups, and obtained two types of compounds with D,L-configuration, as illustrated below:

$$H_2N—\underset{\substack{|\\ CH_2COOH}}{CH}—CO—NH—\underset{\substack{|\\ R^2}}{CH}—CH_2—COOR^1$$

(I)

$$H_2N—\underset{\substack{|\\ CH_2COOH}}{CH}—CONH—CH_2—\underset{\substack{|\\ R^2}}{CH}—COOR^1$$

(II)

$$H_2N—\underset{\substack{|\\ CH_2COOH}}{CH}—CONH—\underset{\substack{|\\ R^2}}{CH}—CH_2OOCR^1$$

(III)

When considering the chain length, some of these compounds compare closely with the L-α-aspartyl-D-alanine *n*-propyl ester which is 170 times sweeter than sucrose. However, large structural changes had an adverse effect on the sweetness.

The structural changes carried out in the compounds of the type III were more successful. The lengthening of the chain by the insertion of the CH_2 group was accompanied by a modification of the terminal carboxyl group. Eleven esters in this series were found to be at least 100 times sweeter than sucrose [169]. The structural requirements of the two hydrophobic moieties R^1 and R^2 were the same as for the dipeptide esters. 2-Aminoalkanol esters must have a configuration that corresponds to the D-form of the amino acids; none of the esters of 2-(α-L-aspartylamino)-L-alkanols was sweet.

The relationship between the structure and taste has been extensively studied for the α-aspartyl derivatives of hydroxyamino acids [167, 191]. The esters of L-α-aspartyl-L-β-*erythro*-hydroxynorleucine, L-α-aspartyl-L-β-*threo*-hydroxynorleucine, *O*-acyl-L-serine, D-serine,

D-threonine and D-allothreonine were synthesized and their sweetness compared with that of the corresponding ester dipeptides without the hydroxyl group [167]. The comparison revealed that the hydroxyl group had a definite effect on the sweetness; its presence in the L,L-series suppressed the sweetness but enhanced it in the L,D-series. Thus the esters of α-L-aspartyl-D-serine and D-threonine were sweeter than the corresponding esters of α-L-aspartyl-D-alanine and D-α-aminobutyric acid. Furthermore, the *erythro-* and *threo*-isomers turned out to have different sweetness potency [167]. The methyl esters were the sweetest amongst the various possible esters of α-L-dipeptides [103]. The *n*-propyl and 2-methylpropyl esters were the sweetest ones in the analogous hydroxy substituted L,D-series [167, 169].

Amongst the 2-α-L-aspartylamino-D-alkanol esters [169], an exchange of methyl group by a hydroxymethyl resulted in an enhancement of sweetness in one case but a decrease in all other cases.

Modifications in the side-chain generated by the introduction of a methylene group in β- or γ-position of the hydroxy dipeptide esters produced bitter compounds [167]. Data on the function of the hydroxyl group for creating a hydrophilic interaction with the receptor exemplify the enormous selectivity of this interaction.

The function of the side chain of α-aspartyl dipeptide esters for eliciting sweet taste have been studied in detail [191]. The authors synthesized a set of structurally related esters of *S*-alkyl-L-cysteine, *O*-alkyl-L-serine, L-serine and unsymmetrical aminomalonic diesters, and determined the dimensions of the side-chain by utilizing the molecular models. Their conclusions [191] were essentially the same as those reached in reference [166]; they stressed that the intensity of all investigated compounds depended on the hydrophobicity, dimensions, shape and stereochemical peculiarities of the side chain. The authors [191] demonstrated convincingly that the knowledge of the degree of modification in the side chain R^2 alone had little predictive power, the overall geometry enabled a more meaningful assessment of the eventual sensory properties.

All the results discussed above indicate that it is the side chain R^2 of the ester dipeptide that is responsible for the interaction with the hydrophobic part of the receptor [192]. However, this postulate applies only to the L,L-dipeptides; in the L,D-series different spatial arrangement of the substituents at the asymmetric carbon atom in the D-amino acid introduces the hydrophobic interactions only at the terminal *C*-ester group, or the OCOR group as in *N*-α-L-aspartyl-D-aminoalkanols This conclusion explains readily an early empirical rule [145, 166, 167, 197] according to which the sweetness of α-aspartyl derivatives with a D,L-configuration is directly proportional to the size of the ester group and inversely proportional to the bulkiness of the side chain.

3.4.3 α-Aspartylamides

Mazur and co-workers [165], having synthesized a series of α-aspartylamides, had demonstrated that an ester group was not essential for the sweetness of the derivatives of L-aspartic acid. Later, other derivatives, other than esters and amides, proved to be also sweet, for example, acylaminoalkanols [198]. In fact, these results were not so surprising since the changes in the carboxylic function of L,L-dipeptides did not affect the part of the molecule that is directly responsible for the interaction with the receptor — glucophore. Although the structure of the amides is not free of peculiarities, essentially the same structural requirements apply to both α-aspartylamides and esters. Fundamentally, the sweetness depends on the presence of a methyl group in α-position.

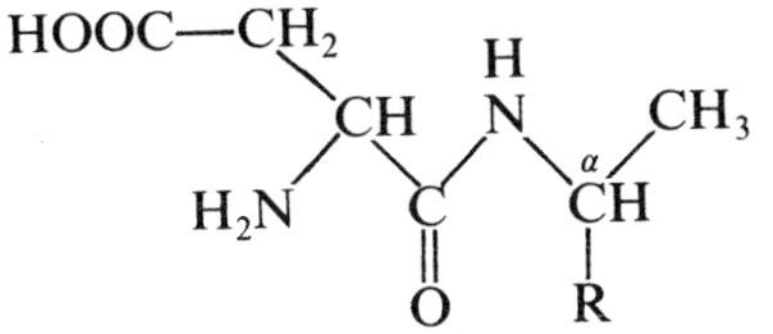

HOOC—CH_2 H
CH N CH_3
α
H_2N C CH
O R

In addition, when optically active amines are utilized as the starting material, the enantiomer that will imitate the spatial arrangement of an L,L-dipeptide should be selected.

An introduction of a halogen in the *p*-position of the phenyl group or its reduction failed to increase the sweetness of α-aspartylamides [165]. As in dipeptide esters, the aromatic moiety can be replaced by an alkyl chain — the optimal length was found to be seven carbon atoms. Either lengthening or shortening of the chain resulted in decreasing sweetness or even tasteless compounds. The sweetness of α-aspartyl-α,δ-dimethylamide is 100 times sweeter than sucrose, but none of the amides were sweeter than aspartame itself. Later, sweet α-aspartyl-D-alanylamides [199] of the type

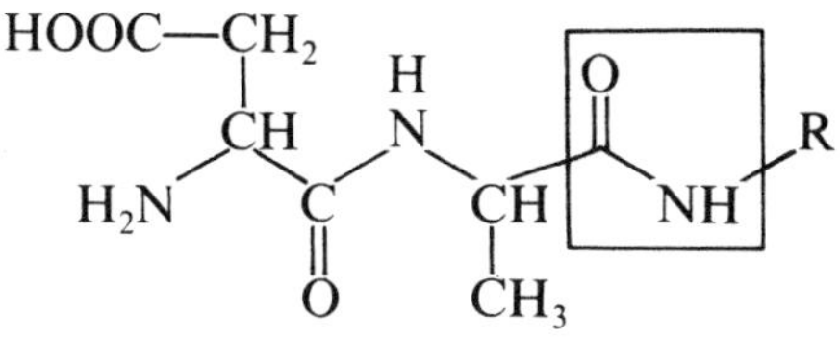

HOOC—CH_2
H O
CH N R 'retro-inversion'
H_2N C CH NH
O CH_3

R = Bu, *i*-Bu, *t*-Bu, cyclohexyl

or amides with the reversed direction of the amide bond, *N*-acetylated *N*-(L-aspartyl)-1,1-diaminoalkanes were synthesized [200–202].

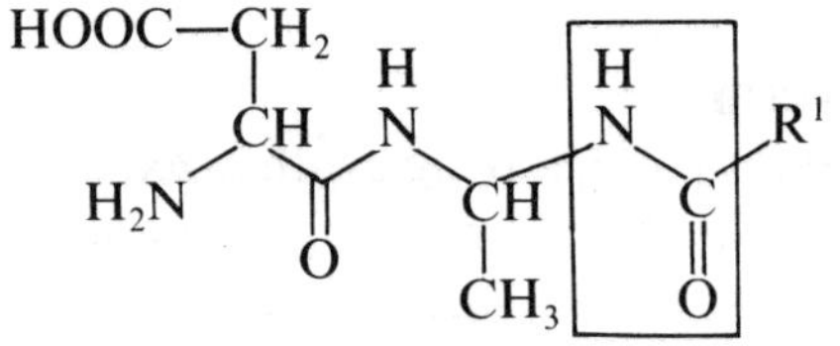

Hofmann's rearrangement is the crucial step in the synthesis of the 1,1-diaminoalkanes; it is facilitated by a mild oxidizing reagent — bis(trifluoroacetoxy)iodobenzene [201] (Scheme 3.30).

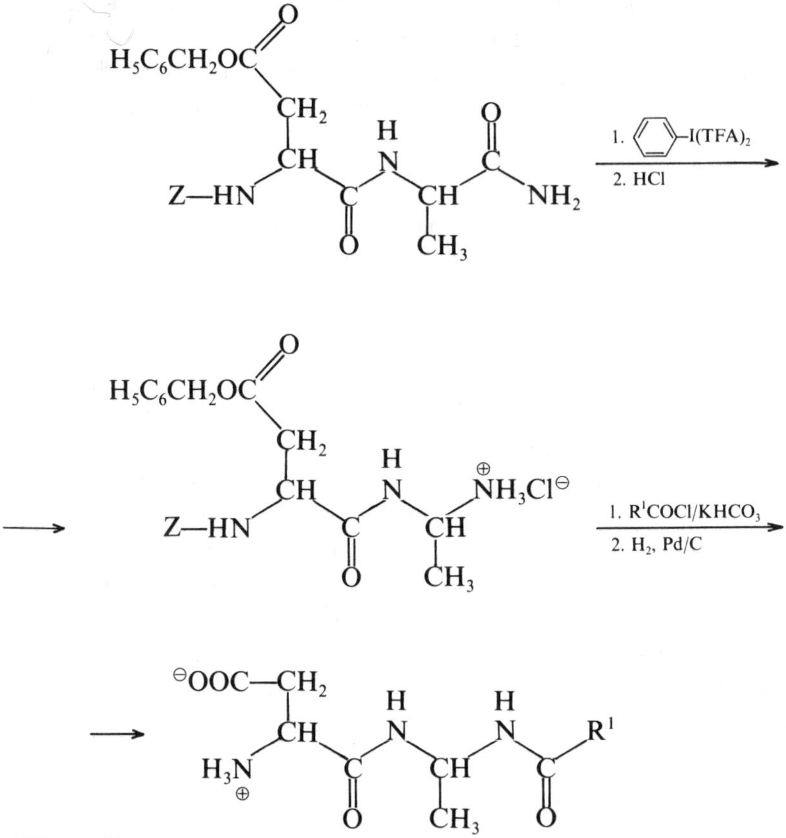

1. C_6H_5–I(TFA)$_2$
2. HCl

1. $R^1COCl/KHCO_3$
2. H_2, Pd/C

TFA = trifluoroacetoxy

Scheme 3.30

The sensory properties of these amides were surprisingly similar to those of sucrose. The intensity of the sweet taste depends on the character of R^1 (acylating carboxylic acid) and varies between 75 and

100 times that of sucrose R = *t*-alkyl or cyclopentyl derivatives are even 800 to 1000 times sweeter (R^1 = 2,2,5,5-tetramethylcyclopentyl). Interestingly, the chirality of 1,1-diaminoalkanes seems to have no relation to sweetness, in contrast to analogous dipeptide esters and amines.

A series of hydrolytically more stable *α*-aspartylamides derived from amino alcohols has been prepared by Rizzi [203] and Rizzi and Echler [204]; their stability was higher than that of aspartame at neutral pH.

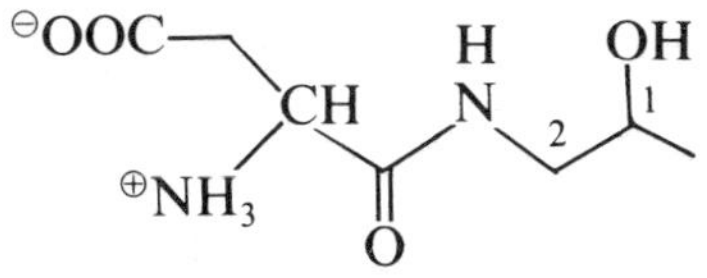

Their sweetness is not outstanding and depends on the configuration and substitution at C_1 and C_2. The sweetest representative of this group was 60 times sweeter than sucrose.

3.4.4 *N*-Acyl-*α*-aspartyl derivatives and their analogues

The fact that some *N*-acyl derivatives of aspartic acid were sweet was rather surprising [205]. The sweetness of one of them, trifluoroacetyl-*α*-L-aspartylanilide was, yet again, discovered by chance [205]. Many analogues were synthesized later in which the *N*-terminal amino acid was replaced either by aspartic, glutamic [209] or aminomalonic acid [210].

The structural attributes essential for the manifestation of sweetness in *N*-trifluoroacetyl-*α*-L-aspartylanilides have been recognized to be [210, 211]:

(a) NH— bond of the anilide must engage in the interaction with the receptor. When it is methylated or its hydrogen atom is involved in other interactions, the compounds are bitter
(b) Substitution in the *m*-position of the aromatic ring changes the shape of the molecule and thus affects adversely the interaction with the hydrophobic part of the receptor. Since the bitterness receptor is apparently more tolerant of such changes in bulkiness, substituted derivatives are bitter [212]
(c) The extrapolation from the observed bitterness of 2-methyl substituted nitroanilides (Table 3.8) indicates that if the anilides are to be sweet, the NH bond and the aromatic ring must be coplanar

It thus appears that bitterness and sweetness in L-aspartates are closely related and only minor structural changes can lead to a change from one to the other.

Table 3.8 — Sensory properties of substituted trifluoroacetyl-α-L-aspartyl anilides

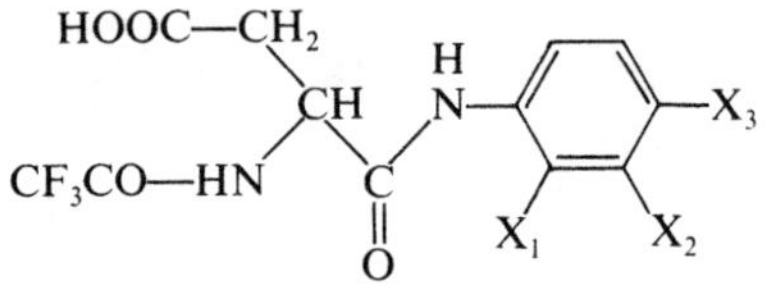

$$HOOC-CH_2-CH(NH-COCF_3)-C(=O)-NH-C_6H_2(X_1)(X_2)(X_3)$$

X_1	X_2	X_3	Taste	References
H	H	H	12	[205]
H	H	Cl	120	[205]
H	H	CN	3000	[205]
H	H	NO_2	1500	[206]
H	H	CF_3	0	[205]
H	H	I	0	[205]
CH_3	NO_2	H	bitter	[210]
Cl	H	NO_2	bitter	[210]
H	CH_3	NO_2	bitter sweet	[210]

3.4.5 Aminomalonic acid esters

The unsymmetrical esters of the α-aspartylaminomalonic acid [163, 213–216] clearly lead the field of synthetic sweet compound because of their sweet potency. The aminomalonic acid itself is unique in as much that its derivatives are potent sweeteners irrespective of its position of attachment, either *via* the *C*- or the *N*-terminal position, of the dipeptide.

The unsubstituted aminomalonic acid lacks chirality, the asymmetry arises in the derivatives when the carboxyl groups are differently derivatized: for example, as an amide and an ester or as a diester with two different alkyl groups. However, the newly created centre of asymmetry is racemized quite readily. Of all the esters of the aminomalonic acid, the methyl ester of D,L-aminomalonyl-L-phenylalanine has the closest resemblance to aspartame but is superior to it in its sweet potency [213–216].

The experimental data [163, 213–216] indicate that the two ester groups of the aminomalonic acid differ in bulkiness (Table 3.9). The sweetest among them possess one methyl ester group, the other is responsible for the hydrophobic interactions with the receptor, and thus for the overall sensory properties of the molecule. It is found that in compounds of the type

Table 3.9 — Derivatives of α-aspartylaminomalonic acid

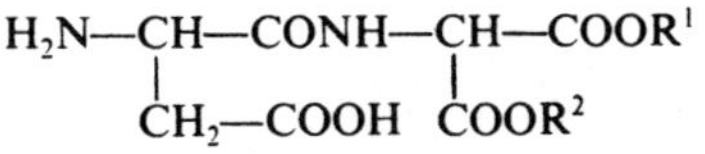

$$H_2N—CH(CH_2—COOH)—CONH—CH(COOR^2)—COOR^1$$

R^1	R^2	Relative sweetness	References
CH_3	*i*-Amyl	500	[215]
CH_3	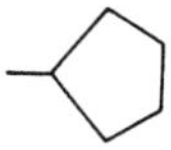	300–600	[163, 215, 217]
CH_3	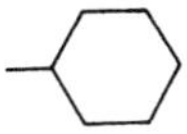	556–880	[163, 215, 217]
CH_3	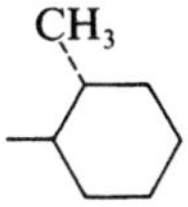 CH_3 *trans*	5450–7300	[163, 215, 216]
CH_3	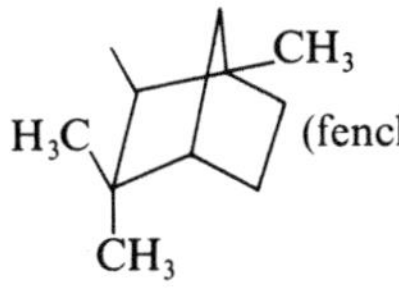 CH_3 H_3C CH_3 (fenchyl)	22,200–33,200	[163, 215]
C_2H_5	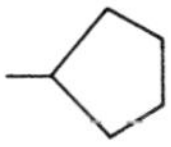	128–156	[215]
C_2H_5	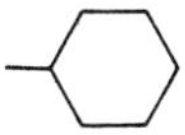	192–284	[215]
C_2H_5	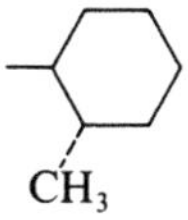 *trans* CH_3	524–648	[215]
C_2H_5	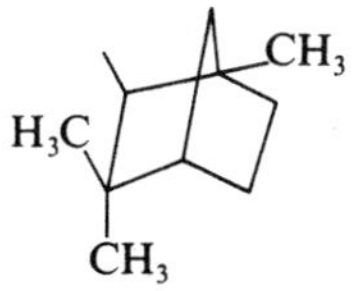 CH_3 H_3C CH_3	4200–5400	[214, 215]

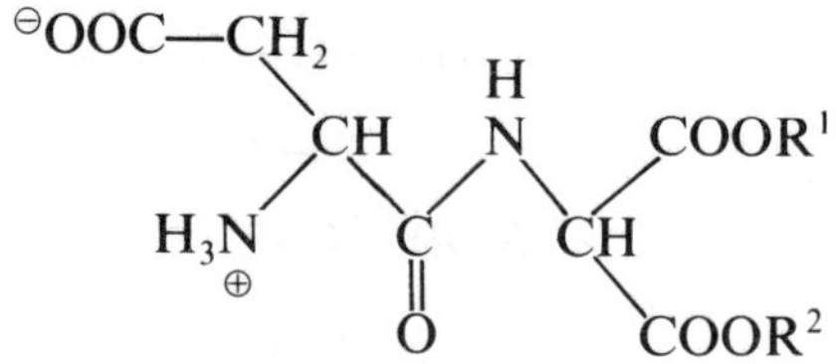

$^{\ominus}OOC—CH_2$
H
CH N $COOR^1$
$H_3N^{\oplus}$ C CH
O $COOR^2$

$R^1 = CH_3, C_2H_5$

the sweet potency rises when R^2 changes from alkyl < cycloalkyl < cycloalkyl substituted in position 2 < fenchyl. When the cyclohexyl group has substituents in another position than 2, or is more than monosubstituted, the sweetness decreases [170].

It can be seen from Table 3.9 that the fenchyl derivative of *α*-aspartylaminomalonic acid is the sweetest compound known. Its sweetness is expected to be even greater because the assayed derivative was not optically pure. Although it is not known which of the diastereoisomers is sweet, Ariyoshi's hypothesis predicts that it should be the one with the *R*-configuration at the *C*-terminal centre of chirality [218].

Fenchyl esters of other dipeptides, that have *α*-alanine or serine in place of aminomalonic acid, aroused great interest due to their outstanding taste properties [219–222].

3.4.6 Aspartame — physico-chemical properties, metabolism and toxicity

Methyl ester of *α*-aspartyl-L-phenylalanine — Aspartame (APM) ($C_{14}H_{18}N_2O_5$, $M = 294.3$) is a white crystalline compound with m.p. 246–247 °C. It can be prepared as a hemihydrate or anhydrous. Its structure was confirmed by X-ray crystallographic studies [223]. Aspartame cyclizes on heating to the corresponding substituted diketopiperazine derivative (DKP) (3,6-dioxo-5-benzyl-2-piperazinylacetic acid) (Scheme 3.31).

The cyclization takes place in aqueous media and is pH- and temperature-dependent. The rate of formation of DKP in neutral and basic conditions is quite high [223, 225]. The epimerization of DKP and the dipeptide *α*-aspartyl-L-phenylalanine (AP) is also strongly pH- and temperature-dependent. Measurements performed under neutral conditions show that the relative rates of racemization of DKP and AP formed a series DKP > NH_2-terminal > COOH-terminal [225].

The dioxopiperazine derivative itself is not only tasteless but its presence also decreases the sweetness of aspartame. Figure 3.3 illustra-

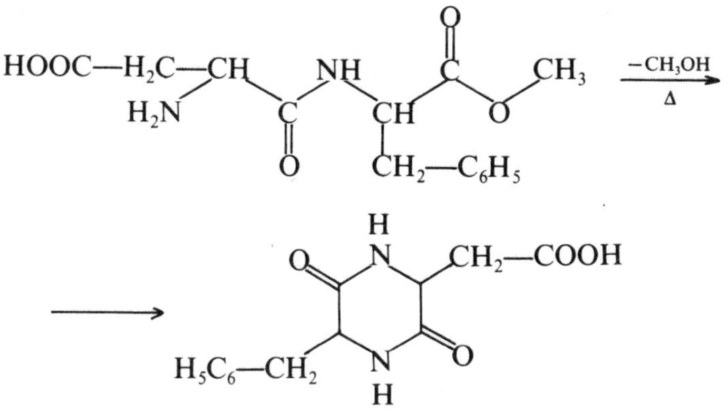

$$HOOC-H_2C-CH(NH_2)-C(=O)-NH-CH(CH_2-C_6H_5)-C(=O)-O-CH_3 \xrightarrow[\Delta]{-CH_3OH}$$

$$\longrightarrow$$ 2,5-dioxopiperazine ring: (H)N, O=, CH_2-COOH, $H_5C_6-CH_2$, N(H), =O

Scheme 3.31

tes the loss of stability of aspartame both in strongly acidic and neutral--to-basic media — aspartame is most stable in the pH range 3.0–5.0, with an optimum at 4.3 (Table 3.10). An isoelectric point of aspartame is reached in 1 % aqueous solution at 20 °C. The maximum solubility (10 %) is reached at pH 2.2; even a 30 % solution of aspartame can be prepared in 70 % acetic acid. Aspartame has a remarkable shelf life — storage in a closed system at 40 °C for one year resulted in a formation of only 1 % dioxopiperazine and 5 % aspartylphenylalanine (by thin-layer chromatography).

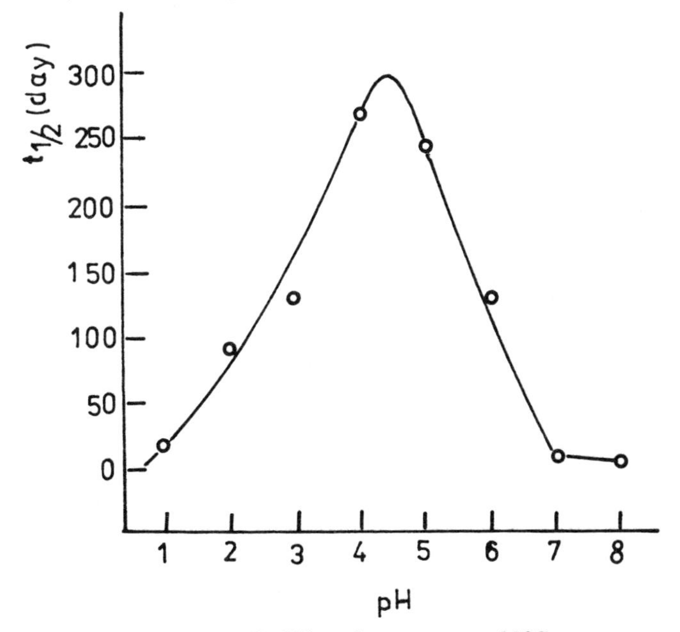

Fig. 3.3 — Stability of aspartame at 25 °C.

Table 3.10 — Stability of aspartame at various pH and temperatures

pH	Temperature (°C)	Half-life (h)
2	40	568
4	40	1644
7	32	10.2
7	40	3.9
8	40	1.4

Stability tests performed under extreme conditions revealed that aspartame remained quite stable at 105 °C, and contained only 5 % of DKP after 75 hours. Aspartame decomposes at temperature greater than 150 °C.

In vivo aspartame is metabolized into its constituents — methanol, phenylalanine and aspartic acid. The hydrolysis of the ester bond gives rise to the parent dipeptide AP which, under certain conditions, exists in an equilibrium with dioxopiperazine. The dipeptide can be hydrolysed further into the individual amino acids (Scheme 3.32).

Aspartame metabolism generates 16.7 kJ/g and produces compounds commonly found in human diet. One of them is an essential

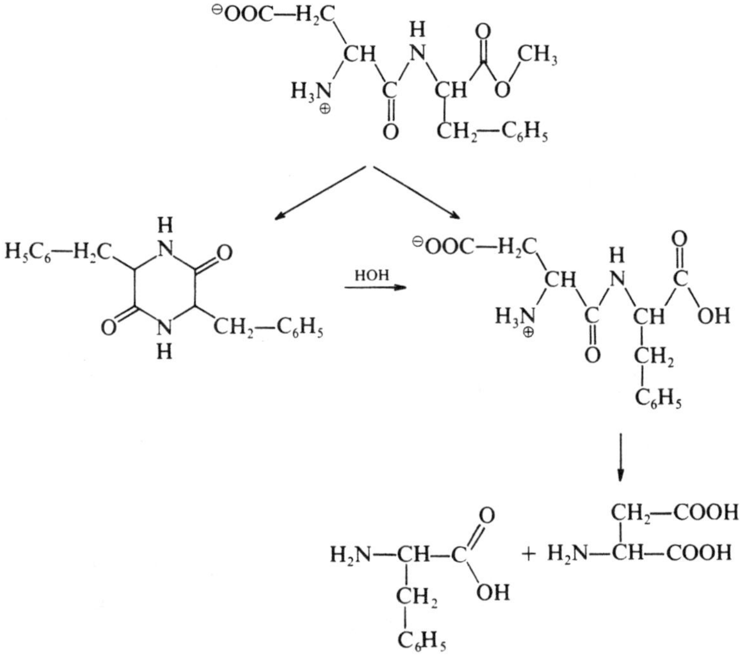

Scheme 3.32

amino acid — phenylalanine, another one, aspartic acid, is present in all proteins. Methanol is also quite common in foods that we consume. Acute toxicity of aspartame is rather low, LD_{50} values for mice are 13.0–16.7 g/kg (per os) and 12 g/kg (intraperitoneal administration), and 5 g/kg (per os) for rats and rabbits. Chronic toxicity was monitored on mice which were given high daily doses up to 4 g/kg of aspartame for as long as two years. The animals showed no symptoms that were significantly different from those in the control group [186].

In order to obtain a better understanding of aspartame metabolites, it is instructive to compare qualitatively a 19 mg dose of aspartame (sufficient to substitute a teaspoonful of sugar) with a glass of milk and tomato juice (250 ml). Whilst the above dose of aspartame produces 7.6 mg of aspartic acid, 9.5 mg of phenylalanine and 1.9 mg of methanol, a glass of milk metabolizes to 528 mg of aspartic acid and 542 mg of phenylalanine (60 times the amount produced by aspartame). A glass of tomato juice produces, after being metabolized, 47 mg of methanol, an equivalent to 25 glasses of drink sweetened by aspartame.

Aspartame was approved as a sweetener and a flavour enhancer by the U.S. Food and Drug Administration (FDA) for table use and as an ingredient in certain food applications on 25th July 1974. The regulation became codified in CFR 172.804.

Formal objections to the regulation were filed immediately after the approval: they were based on safety connected with the toxic effects as aspartame metabolites — phenylalanine, aspartic acid and substituted dioxopiperazine (diketopiperazine, DKP). The regulation was suspended in December 1975, pending further findings and a Public Board of Inquiry was established to examine the objections. Finally, after five years, the Commissioner of Food and Drug Administration reviewed in detail the contested aspects of aspartame safety and found that aspartame had been shown to be safe, and vacated the stay of the original regulation (October 1981).

More than 100 studies have been undertaken with the aim to identify the possible health hazards associated with aspartame metabolites. Both normal (lower than 36 mg/kg/day) and abnormal daily doses were used. Although all aspartame metabolites, with the exception of DKP, are a part of human diet, either as a direct constituent of food or arising during food processing, their effect can be still harmful, especially if present in larger amounts. These fears fuelled further detailed studies of aspartame metabolites.

Aspartic acid metabolizes into glutamic acid. Since a high level of any of these amino acids is potentially dangerous to the nervous system, their level in blood was correlated with normal and abnormal intake of aspartame. The result was negative both for adults as well as for children [226–228].

Persons born with phenylketonuria cannot metabolize phenylalanine; they lack the enzyme that catalyses the transformation of phenylalanine into tyrosine. High doses of accumulated phenylalanine damage the brain at an early age. Sweeteners containing aspartame are contraindicated for such persons.

High doses of aspartame (250 mg/kg) administered to rats led to an increased level of phenylalanine in the brain and in plasma [229, 230]; the level of phenylalanine in blood remained in humans [231]. Extensive studies on normal adults, children, pregnant women, foetuses, lactating women, sucklings and diabetics have proved aspartame to be safe [226, 228, 232]. When the FDA reviewed the results of the above studies, it reached the conclusion that the use of aspartame posed no danger of brain damage due to an increased phenylalanine level in blood plasma (46 FR 38291).

A number of unanswered questions remained concerning the role of increased phenylalanine concentration in the blood amino acids, from which the neurotransmitters are produced. If, in any way, phenylalanine could affect the production and concentration of neurotransmitters, it would also affect the brain functions, and ultimately the behaviour of the person. Careful studies were carried out on test animals which were fed with normal and abnormal doses of aspartame and it was found that neither the neurotransmitter levels, behavioural pattern and brain waves nor physical and mental development were significantly affected [233–235].

Studies on adults and children, carried out over a period of 21 weeks, who were given daily doses of 34–200 mg/kg (or 1800 mg/kg) of aspartame, failed to reveal any detectable changes in the behaviour of the subjects. However, high doses of aspartame fed to rats affected their amino acid concentrations in the brain with the concomitant changes in neurotransmitter levels [229]. Cases of abnormal behaviour, after high consumption of sweetened drinks have been known even in humans [236]; this area still remains the focus of interest of the investigators [237].

Methanol metabolizes to formaldehyde, which in turn is quickly oxidized to the formate ion. Thus the toxic level of the formate represents a measure of toxicity of methanol. Studies undertaken with the aim of clarifying this point had failed to detect methanol concentrations that would have caused formate toxicity. This led the FDA to state that there was no reason to fear toxicity on account of methanol derived from a higher intake of aspartame.

Dioxopiperazines (DKP) are intermediary metabolites of aspartame and they arise from it on heating or after longer storage at elevated temperatures. Animal tests failed to reveal any muta-, carcino- or teratogenicity [235]. It thus appears that if food that contains aspartame is heated, no harmful products are formed, only the sweetness decreased.

FDA established a special Food Programme for monitoring any harmful effects associated with the use of aspartame in view of the many controversial issues. Special attention is given to certain age groups which are likely to be exposed to higher doses of the sweetener (2–3 years old children). Occasional complaints of aspartame users relating to higher blood pressure, epileptic fits and memory loss are unique, and have no connection with aspartame.

3.4.7 Applications

Aspartame provides a clean, sweet taste that resembles sucrose, without any metallic aftertaste. Products sweetened with aspartame have lower calorific value (Table 3.11). Aspartame has been marketed under the

Table 3.11 — Comparison of energy content of foods sweetened alternatively by aspartame or sucrose [238] (kJ per serving)

Food product	With sucrose	With aspartame
Dry mix lemonade (25 g)	359	21
Dry mix desert (1/2 glass)	338	42
Chocolate pudding from dry mix (1/2 glass)	627	313
Dry mix chocolate drink (25 g)	485	263
Whipped cream (spoonful)	58	21
Instant cream cocktail (1 glass)	790	292

following brand names — Nutrasweet (flavour enhancer) and tabletop sweetener — Equal, Usal, Cauderel, Pouss-suc, D-Sucril, Tri-sweet.

Its sweetness potency is about 160–200 times that of sucrose, however it varies with the type of food it is used with, and with the concentration (Table 3.12).

In general, the concentration of aspartame varies from about 0.01 % when used as a flavour enhancer, to about 0.6 % when used as a sweetener. Its actual concentration level would depend, in each case, on the product type, formulation, pH, temperature and other specific factors.

Aspartame enhances and extends sensory properties of food and drinks, particularly the acidic fruit flavour. Although its sweetness develops gradually, it persists rather longer than that of sucrose. Any

Table 3.12 — Sweet potency of aspartame in various systems

System	Sweet potency[a]	Sucrose (%)	Temperature
Water	400	0.34	22
	215	4.3	22
	133	10	22
	100	15	22
	182	2	8
	160	4	8
	100	12	8
	43	30	8
Citrate buffer	160	4	8
Non-alcoholic dry-mixes	133	10	8
	143	10	3
	180	11	8
Chewing gum	100–199	69	22
Gelatin	160–172	18	3
Pudding mixtures	200	17	8
Non-alcoholic beverages	180	10	8
Tea, coffee	180	4–6	70
Flavoured yoghurt	175–220	17	8
Canned fruit	175	20	cooled

[a] Potency is equal to the concentration of sucrose divided by the concentration of aspartame at equal sweetness.

lingering sweetness of aspartame can be modified by the addition of other sweeteners or salts.

An appreciable reduction in the package size and weight can be achieved by using aspartame; apart from the economic benefits, one also obtains a more favourable ratio of ingredients and more nutritious components. For example, if some or all of the sugar in a presweetened cereal is replaced with aspartame sweetness, the amount of grain in the packet can be increased by 10–40 %. A dry-mix beverage which makes 2 litres of the final product weighs 200 g when made with sucrose and only 14 g when made with aspartame; this represents a 93 % reduction in net weight, a significant reduction in the cost of packaging and shipping.

Aspartame exhibits synergy with many sweeteners thereby allowing a reduction in the level of the total sweetener (see also p. 215). Aspartame is used as a tabletop sweetener, in combination with diluents, in the form of tablets, granules or in sachets. Other applications include dry mix products, carbonated beverages, cereals, chewing gum, dairy products, frozen drinks, confections, applications in pickling, creme

Table 3.13 — Concentration of aspartame in carbonated drinks

Drink	pH range	Aspartame concentration ($\% \times 10^{-2}$)
Cola	2.4–3.1	5.5–6.8
Lemon-lime	3.0–3.1	3.0–6.0
Orangeade	3.1–3.4	5.5–9.0
Root beer	4.0–4.4	5.5–7.0

fillings, glazes, jam jellies, preserves and fruit fillings. Table 3.13 lists the typical concentration level of aspartame in beverages.

3.4.8 Determination of *α*-aspartyl-L-phenylalanine in food preparations

Spectroscopic and fluorometric methods have been employed, mainly for the determination of aspartame in simple materials, e.g. for the determination of its purity or its concentration in drug formulations, dry mix beverages or tablets.

UV spectrophotometric determination of aspartame has been based on its ability to complex with copper sulphate [239]. The complex is isolated by extraction and the copper content is determined as the diethyldithiocarbamate salt with a maximum at 436 nm. The method is unreliable if other compounds are present that form complexes with Cu^{2+} ions. A colour reaction of aspartame with *p*-dimethylaminobenzaldehyde was used by Özol [240]. Another quantitative determination of aspartame is based on a colour reaction with 1,4-benzoquinone [249].

Quantitative determination of aspartame as a pure substance or in tablet preparations was published by Počtová and Kakač [242] — it utilizes the fluorescence of the reaction product of aspartame with fluorescamine, which is formed at room temperature and pH 7.5. The fluorescence intensity (both at excitation and emission wavelengths of 309 and 480 nm, respectively) correlates linearly with the concentration of aspartame in the range of 0.25–1.0 μg/ml.

Chromatographic methods have been the most popular choice for the determination of artificial sweeteners. Japanese workers detected aspartame in an assortment of drinks by thin-layer chromatography on cellulose [243]; they used 2-butanone–pyridine–acetic acid–water as the eluent. A similar routine analytical method for aspartame was developed in the soft drinks industry [244]; it is applicable mainly to the analyses of syrups and soft drinks. The detectable threshold of aspartame by ninhydrin on cellulose layers Lucefol Quick, developed by a mixture of 2-butanone–pyridine–water (14 : 3 : 3) was 0.1 μg. Thin-layer chromatography was also employed for the analysis of fruit juices [245]. The authors used commercial Whatmann LHPKD (high resolution silica gel) 10 × 20 cm precoated plates and *n*-butanol–acetic acid–water (6 : 2 : 2) eluent. Again, the detection was carried out by spraying the plates by a solution of 0.2 % ninhydrin in acetone and assessed by a densitometer.

Impurities and by-products were detected in the same way by thin-layer chromatography on the gypsum bound silica gel plates [246]. The compounds that could be identified by this method are *α*-aspartame, *β*-aspartame, formylaspartame, 5-benzyl-3,6-dioxo-2-piperazinylacetic acid, phenylalanine methyl ester, aspartylphenylalanine, phenylalanine and aspartic acid. A mixture of 2-butanone–acetic acid–water (70 : 15 : 15) was used as the mobile phase and the detection was carried out by a 1 % ethanolic ninhydrin solution.

Numerous papers have been published on the analysis of aspartame in drinks and tablets [247–249].

Liquid chromatography has been successfully utilized in the analysis of aspartame in blood plasma. On the other hand, Fletcher and Moore [250] utilized an automatic amino acid analyser.

Japanese authors [251] simplified the aspartame analysis further by using only a short column in the amino acid analyser (150 × 9 mm, only half full), eluting with a citrate buffer (pH 4.25) containing 1 % benzyl alcohol at 55 °C. This method was modified further [252, 253]. Aspartame was also assayed by using a combination of amino acid analyser and a reversed phase HPLC [254].

High-pressure liquid chromatography (HPLC) has established itself as a simple but powerful tool for aspartame assay [243, 255]. Several authors [246, 256] deployed this method for the analysis of the impurities in aspartame, for example, the determination of dioxopiperazine, aspartyl phenylalanine and formyl aspartame.

HPLC was also successful in separating aspartame from its degradation products such as *α*-L-aspartyl-L-phenylalanine and dioxopiperazine, synthetic intermediates and by-products such as *N*-formyl-*β*-aspartame and *β*-aspartame [257]. The separation was performed on a 120 × 4 mm Hypersil MOS column with a mixture of acetonitrile, potassium dihydrogen phosphate and phosphoric acid as the mobile phase. Degradation products of Usal were analysed in model systems and in non-alcoholic beverages [258]. The analysis confirmed the presence of aspartic acid, phenylalanine, aspartyl phenylalanine, phenylalanylaspartic acid and dioxopiperazine, apart from Usal itself. The build-up of dioxopiperazine in soft drinks during extended storage received special attention. A gradient method was devised in order to isolate the hydrolytic and degradation products of aspartame and detect them by means of UV spectroscopic and fluorescence methods [259].

A fast method for aspartame assay in solid and liquid food preparations with higher fat content has been developed [260]. Food samples were first extracted with water, then phosphate buffer (pH 4) and cetyltrimethylammonium chloride were added, and the solution was pumped through a SEP-PACK C18 column. The column was then washed with water and aspartame eluted with methanol–0.007 *M* phosphate buffer (pH 5) (1 : 3). The quantitative determination was carried out on a Lichrosorb RP-8 column eluted with methanol–phosphate buffer (see above) (1 : 4) and attached to a UV detector set at 210 nm. Employing these conditions, the yield of aspartame from liquid samples was 96.9–100.0 %, from solid samples 91.1–96.5 % and from food products containing fat was still as high as 85.7–98.0 %.

Aspartame content of the tablet form, which utilizes UV detection, has been described [261].

Aspartame content in Coca-Cola and in the tabletop sweetener Equal was determined by chromatography on a column filled with silica gel impregnated with β-cyclodextrin [262].

Several authors have endeavoured to develop a fast and reliable HPLC based analytical method for the determination of aspartame, its degradation products and impurities [263–269].

Gas-liquid chromatography (GLC) has proved to be also a viable method, provided that the samples can be converted into volatile trimethylsilyl derivatives. Although this idea proved to be very successful specially in the pre-HPLC period, it never became too popular. The biggest handicap in this method is the necessity to search for suitable derivatizing agents and the corresponding chemical methodology.

Usal and its degradation products were semiqualitatively assayed in lemonades and syrups by high-voltage electrophoresis on Whatman chromatography paper [270].

Capillary isotachophoresis was first deployed for the analysis of impurities in Usal by Boček [271]. Isotachophoresis in an anionic cycle, coupled with conductometric detector, was used by Kvasnička [272] to evaluate the Usal content in food articles, non-alcoholic beverages, coffee and yoghurt. The method can also be used for the purity control of the sweetener itself, as well as for the monitoring of intermediates during the synthesis.

A novel approach of aspartame assay has been developed by German workers who utilized an amperometric sensor based on immobilized *Bacillus subtilis* cells [273]. The sensitivity of the sensor to aspartame was an order of magnitude greater than to its constituent amino acids. The sensor was stable for eight weeks.

Another sensor was made by chemically immobilizing L-aspartase on an ammonia-selective electrode; it was used successfully for aspartame assay in commercial sweeteners [274].

As can be seen from the above review, there are numerous methods for aspartame detection.

3.5 DERIVATIVES OF UREA

Berlinerblau [275] described 4-ethoxyphenylurea as a compound with sweet taste as early as 1884. He synthesized it by treating *p*-phenetidine with phosgene, followed by a hydrolysis with ammonia — the sweet compound that is obtained is known as Dulcin [276, 277].

$$C_2H_5O-C_6H_4-NH-\overset{\overset{\displaystyle O}{\|}}{C}-NH_2 \qquad C_9H_{12}N_2O_2, \qquad M = 180.2$$

Dulcin

Dulcin is a pale yellow crystalline compound with a melting point of 174 °C; it is 250 times sweeter than saccharin and its sweet taste is more agreeable (Table 3.14).

Dulcin is formed by the condensation of *p*-phenetidine with urea at 160 °C [278]:

$$C_2H_5O-C_6H_4-NH_2 + O{=}C(NH_2)_2 \xrightarrow[CH_3COOH]{HCl} C_2H_5O-C_6H_4-NHCONH_2 + NH_4Cl$$

It forms colourless shiny needles, soluble in ethanol and water. Dulcin decomposes on prolonged heating in acidic conditions into non-sweet compounds.

Dulcin has been known under the following brand names: Glycosine, Sucrol, Valzin, Phenetocarbamide [25].

Dulcin analogues were prepared by varying some of its structural

Table 3.14 — Relative sweetness of urea derivatives

$$R^1\text{—NH—}\underset{\underset{\displaystyle O}{\|}}{C}\text{—NH—}R^2$$

R^1	R^2	Relative sweetness	References
$-C_6H_4-OCH_3$	H	< 250	[190]
$-C_6H_4-OC_2H_5$	H	250	[23, 25, 218, 290]
$-CH{=}CH-C_6H_4-OC_2H_5$	H	sweet	[19, 282]
$-C_6H_4-NO_2$	H	tasteless	[285]
$-C_6H_4-NO_2$	CH_2COO^-	10	[285]
$-C_6H_4-NO_2$	$(CH_2)_2COOH$	350	[23, 25, 218]
$-C_6H_4-NO_2$	$(CH_2)_2COO^{-a}$	173, 700	[19, 285]
$-C_6H_4-NO_2$	$(CH_2)_2SO_3^-$	tasteless	[285]
$-C_6H_4-NO_2$	$(CH_2)_2COOCH_3$	tasteless	[285]
$-C_6H_4-NO_2$	$(CH_2)_3COO^-$	tasteless	[285]
$-C_6H_4-CN$	$(CH_2)_2COO^-$	450	[285]
$-C_6H_4-Cl$	$-CH_2COOH$	sweet	[19]

Table 3.14 (continued)

R^1	R^2	Relative sweetness	References
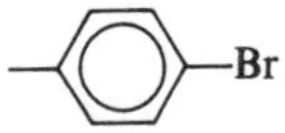 –Br	$-(CH_2)_2COOH$	sweet	[19]
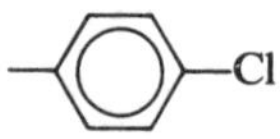 –Cl	$-CH_2COOH$	sweet	[19]
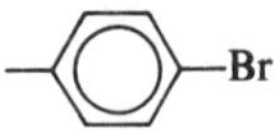 –Br	$-(CH_2)_2COOH$	30	[19]

[a] Thiourea derivative has sweetness 2400 [28].

elements: the *o*-derivative is tasteless and the thiourea derivative is bitter [23]

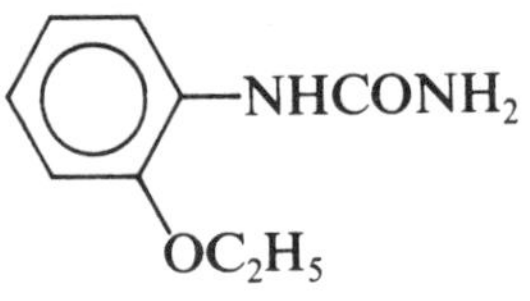

$-NHCONH_2$
OC_2H_5

tasteless

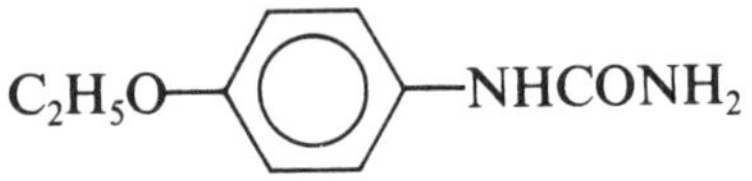

C_2H_5O- $-NHCONH_2$

150–250 times sweeter than sucrose

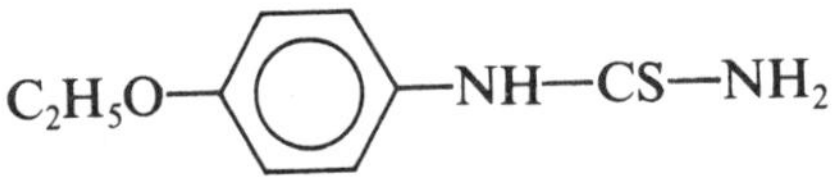

C_2H_5O- $-NH-CS-NH_2$

bitter

The human gastrointestinal tract accepts Dulcin well and metabolizes it completely. Dulcin's toxicity is higher than that of saccharin; it causes cirrhosis in rats. When 50–100 mg/kg of body weight doses are administered over prolonged periods, growth retardation, anaemia and higher death rate result [14]. Because of its high toxicity (dog/per os LD_{50} of 1 g/kg) and suspected carcinogenicity [1], Dulcin has been banned from use in most countries [25, 280].

When still in use, it was employed in combination with saccharin or with sucrose for sweetening of carbonated drinks and in canned fish [25].

Dulcin content in drinks and ice cream was usually determined by colorimetry, fluorometry, oxidometry and by UV spectrophotometry

[278]. The detection in soft drinks could often be performed by thin-layer chromatography [36, 281].

A vinylogue of Dulcin has been prepared and characterized — 4-ethoxystyrylurea [279]. It forms pale yellow needles, m.p. 129 °C, soluble in ethanol. It can be prepared from 4-ethoxybenzaldehyde [282]

$$C_2H_5O-C_6H_4-CH{=}CH-NH-\overset{\displaystyle O}{\overset{\|}{C}}-NH_2$$

$C_{11}H_{14}N_2O_2$ $M = 206.1$

Another sweet derivative of urea is *N*-(4-nitrophenyl)-*N'*-(2-carboxyethyl)urea 'Suosan' [218], accidentally discovered in the study of diisocyanates [283]

$$O_2N-C_6H_4-NH-\overset{\displaystyle O}{\overset{\|}{C}}-NH-CH_2-CH_2-COOH$$

$C_{10}H_{11}N_3O_5$ $M = 253.1$

Pale yellow crystals of Suosan melt at 180 °C, dissolve in water and are stable up to 240 °C. They are also 350 times sweeter than sucrose. Möhler [284] described its structure, properties, reactions and synthesis. The simplest method of preparation is based on the reaction of 4-nitrophenylisocyanate with *β*-alanine:

$$O_2N-C_6H_4-NCO + H_2N-CH_2-CH_2COOH \longrightarrow$$

$$\longrightarrow O_2N-C_6H_4-NHCONHCH_2CH_2COOH$$

As other urea-based sweeteners, Suosan is toxic and excessive amounts taste bitter. It often causes skin rashes and upsets the stomach [286]. Intraperitoneal toxicity in rats, expressed as LD_{50}, was 1 g/kg and thus disqualified Suosan from commercial usage [25].

Esterification with diazomethane yields a tasteless methyl ester [285]. The sweet taste also disappears when the carbon chain is lengthened or if the COO^- is substituted by SO_3^- (Table 3.14).

3.6 TRYPTOPHAN DERIVATIVES

Tryptophan is an essential amino acid, the D-form of which is 25–50 times sweeter than sucrose. The L-form is bitter [19]. The discovery that some tryptophan derivatives are also sweet dates back to 1968. For instance, D,L-6-(trifluoromethyl)tryptophan is an intensely sweet compound [23]. The discovery stimulated the synthesis of other analogues with different substituents in position 6. The preparation usually employed either the Reimer–Tiemann or Mannich reactions, starting with the corresponding indole derivative (Scheme 3.33).

A novel synthesis of racemic 6-substituted tryptophan derivatives

D,L-6-chlorotryptophan

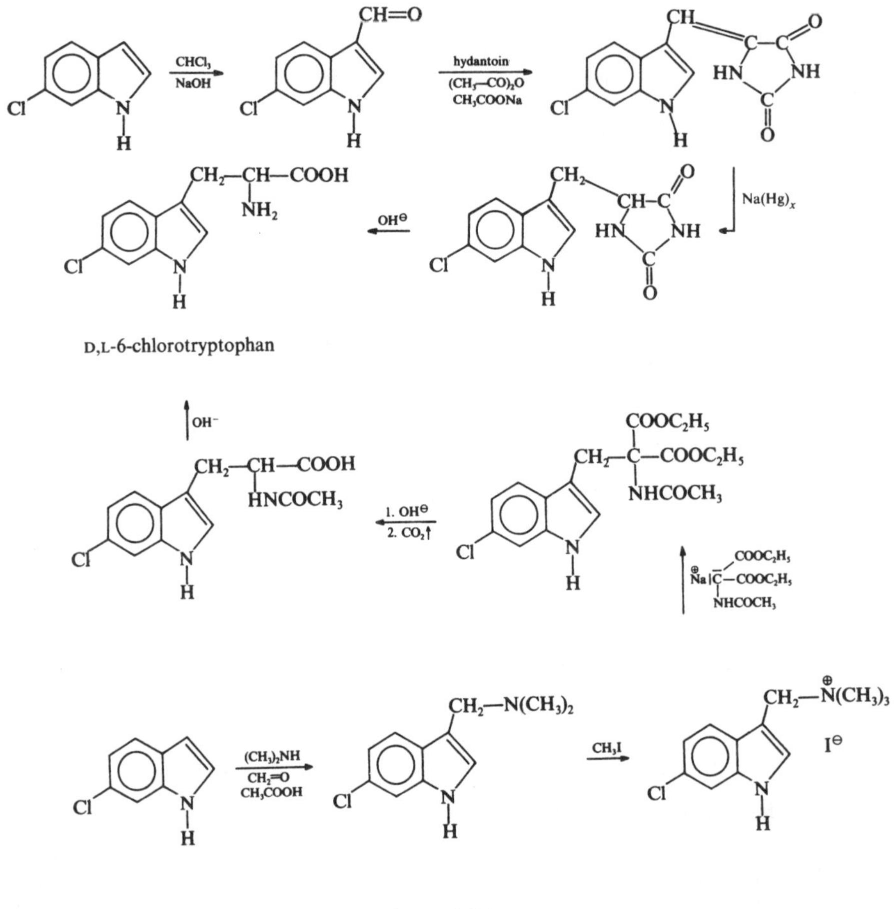

Scheme 3.33

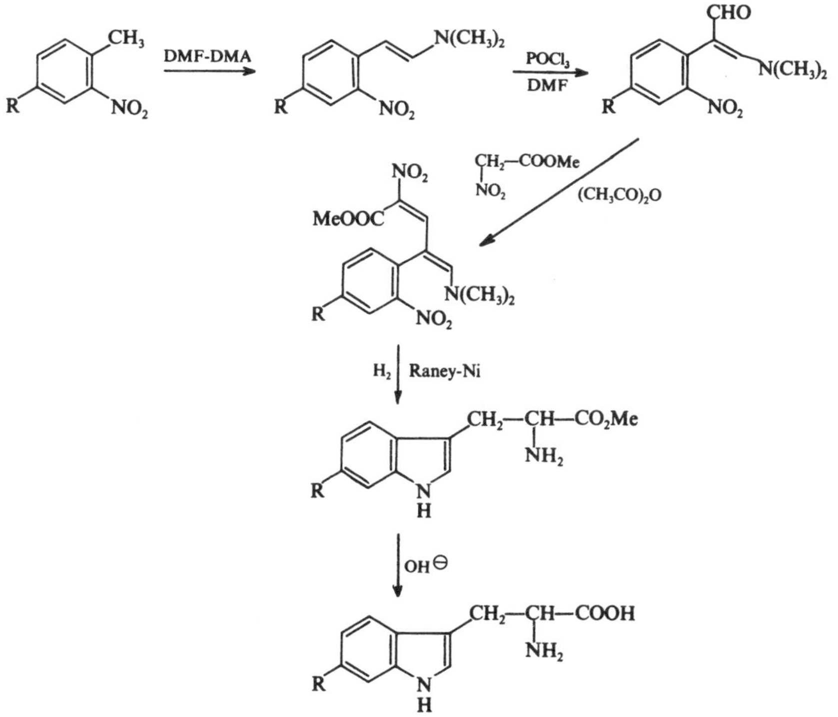

R = CH_3, Cl

DMF-DMA = *N,N*-dimethylformamide dimethyl acetal

Scheme 3.34

was announced in 1979 — it starts with suitably substituted *o*-nitrotoluenes [287] (Scheme 3.34).

During the studies in the indole series, it was found that D-6-chlorotryptophan was 1000–1300 times sweeter than sucrose and free of any aftertaste, whilst the L-form was tasteless but with antidepressant effects [23]. Understandably, the racemic mixtures had to be separated, either by enzymes or by alkaline hydrolysis.

Bromo-, fluoro-, methoxy- and isopropyl-substituted tryptophans were prepared, but nothing is known of their eventual application nor are the results of the toxicological tests.

Cynurenine is a metabolite of tryptophan. Literature describes *N*-formyl- and *N*-acetylcynurenine as sweet substances [288]:

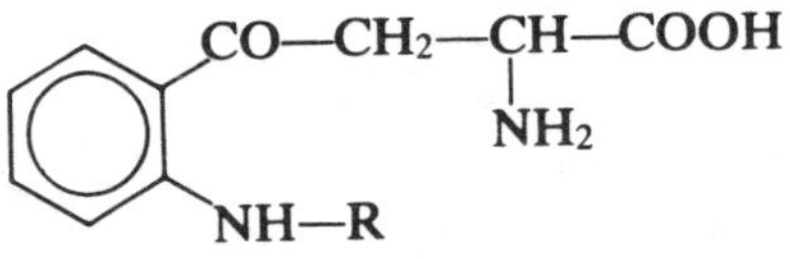

CO—CH_2—CH—COOH
NH₂ → NH_2
NH—R

R = H, CH = O, CH_3CO

The synthesis of cynurenine itself and its formyl derivative was described in 1952 by Dalgliesh [289]. Another synthetic procedure starts with *o*-chloroaniline (Scheme 3.35).

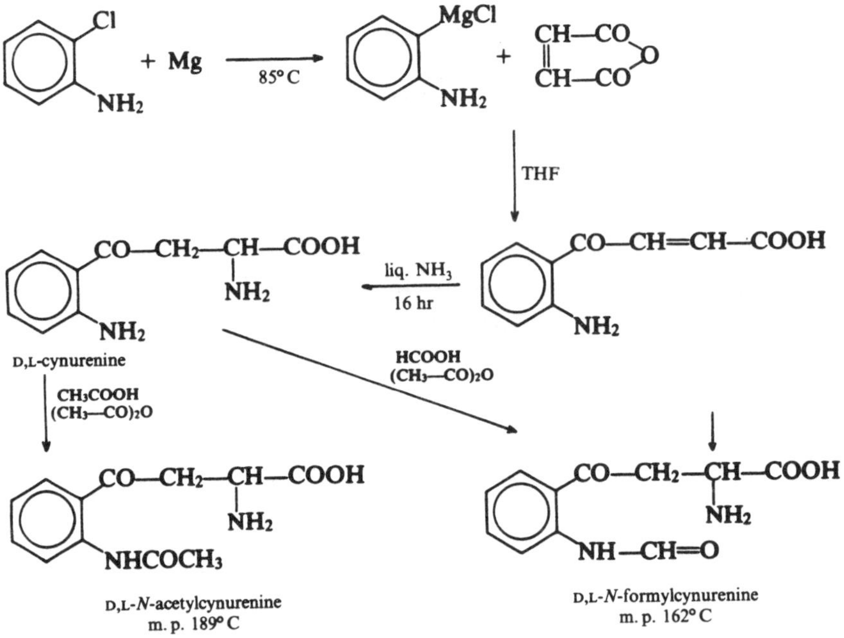

Cl, NH_2 + Mg —(85° C)→ MgCl, NH_2 + CH—CO, CH—CO, O

THF

CO—CH=CH—COOH, NH_2

liq. NH_3, 16 hr

CO—CH_2—CH—COOH, NH_2, NH_2

D,L-cynurenine

CH_3COOH, $(CH_3—CO)_2O$

HCOOH, $(CH_3—CO)_2O$

CO—CH_2—CH—COOH, NH_2, $NHCOCH_3$

D,L-*N*-acetylcynurenine
m. p. 189° C

CO—CH_2—CH—COOH, NH_2, NH—CH=O

D,L-*N*-formylcynurenine
m. p. 162° C

Scheme 3.35

The structures of the compounds were confirmed by ^{1}H NMR and UV spectra. They elicit upon contact with the tongue an immediate sweet sensation; both *N*-formyl and *N*-acetyl derivatives possess 35 times the sweetness of sucrose. No data are available on their applications in food or on their toxicity.

3.7 DERIVATIVES OF BENZENE

World War II was responsible for a greatly increased activity in research and development in the chemical industry, including the field of artificial sweeteners. Many new compounds emerged which had only a very ephemeral existence. The derivatives of 3-nitroaniline belong to this class of sweeteners (Table 3.15).

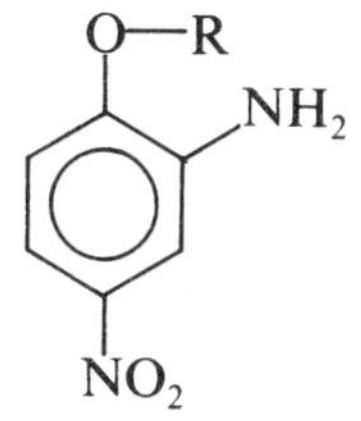

O—R, NH_2, NO_2

$R = CH_3$, C_2H_5, C_3H_7 (*n*- and *iso*-), C_4H_9

The maximum sweetness was reached in the 1-*n*-propoxy derivative (designated also as P-4000) [291], 4000 times sweeter than sucrose and without any bitter aftertaste (Table 3.15). However, its use was often accompanied by skin rashes and stomach disorders [10, 286].

Nitroanilines were synthesized by the Williamson reaction, starting with the corresponding alcoholate and 2,4-dinitrochlorobenzene, followed by a partial reduction of the dinitro derivative [292] (Scheme 3.36 for *n*-propoxy compound).

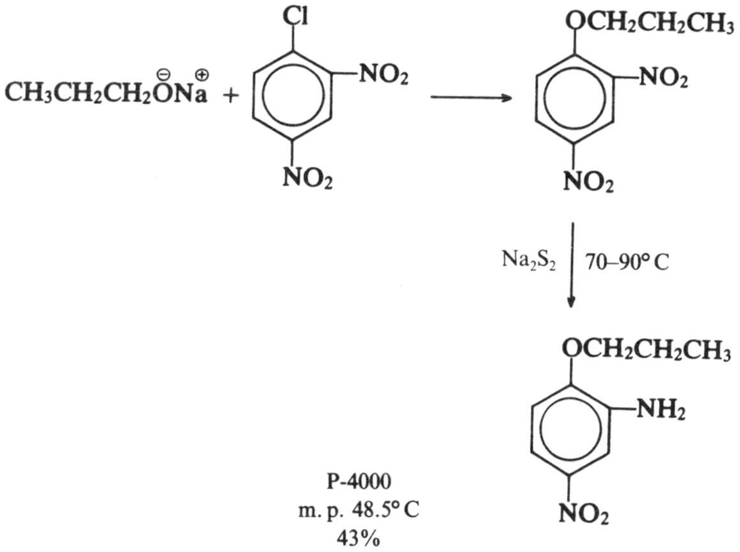

$CH_3CH_2CH_2O^{\ominus}Na^{\oplus}$ + Cl, NO_2, NO_2 ⟶ $OCH_2CH_2CH_3$, NO_2, NO_2

Na_2S_2 | 70–90° C

$OCH_2CH_2CH_3$, NH_2, NO_2

P-4000
m. p. 48.5° C
43%

Scheme 3.36

Duoxane (1-ethoxy-2-amino-4-nitrobenzene) forms yellow, sweet (330 times than sucrose), water and alcohol soluble crystals with m.p. 96 °C (Table 3.15). All sweet compounds in this series are highly toxic, thermally unstable and lack practical importance [286].

Table 3.15 — Relative sweetness of some benzenoid derivatives[a]

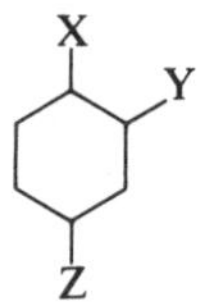

X	Y	Z	Relative sweetness	References
H	NH_2	NO_2	16, 40	[19, 293]
OH	NH_2	NO_2	100, 120	[291, 290]
OCH_3	NH_2	NO_2	108, 160, 220	[19, 286, 291, 293]
OC_2H_5	NH_2	NO_2	330, 506, 950	Duoxan [286, 291, 294]
O-n-C_3H_7	NH_2	NO_2	2350, 4000, 4100	[19, 218, 286, 291, 292, 294, 295]
O-iso-C_3H_7	NH_2	NO_2	500–1000, 600	[19, 286, 291]
O-n-C_4H_9	NH_2	NO_2	614, 1000	[19]
O-sec-C_4H_9	NH_2	NO_2	400, 300–500	[19, 286, 291]
O-CH_2—C_6H_5	NH_2	NO_2	tasteless	[277]
CH_3	NH_2	NO_2	300, 314	[19, 218, 286]
O-n-C_3H_7	NH_2	NO_2	2000	[19]
F	NH_2	NO_2	40	[19, 293]
Cl	NH_2	NO_2	400, 375	[19, 293]
Br	NH_2	NO_2	700, 715, 800	[19, 290, 293]
I	NH_2	NO_2	1250	[19]
CH_2OH	NH_2	NO_2	tasteless	[19, 290]
COOH	NH_2	NO_2	25	[19]
NH_2	NH_2	NO_2	tasteless	[290]
O-n-C_3H_7	NH_2	CN	2500	[290, 296]
C_2H_5	OH	NO_2	sweet	[290]
Br	OH	NO_2	sweet	[290]
$(CH_2)_3COOH$	NH_2	NO_2	tasteless	[290]
O-n-C_3H_7	OH	NO_2	tasteless	[290]
F, Cl, Br	NH_2	CF_3	sweet	biting aftertaste [290]
Cl	OH	CF_3	sweet	phenolic aroma [290]
CH_3	NO_2	NH_2	tasteless	[293]
CH_3O	NO_2	NH_2	tasteless	[293]
CH_3O	OH	$OCOC_2H_5$	sweet	[297]
CH_3O	OH	OCO-2-furyl	sweet	[298]

[a] Related to sucrose as standard.

Attempts to correlate chemical structure and sweetness revealed that trisubstituted benzenes (with substituents X, Y, Z) were sweet when

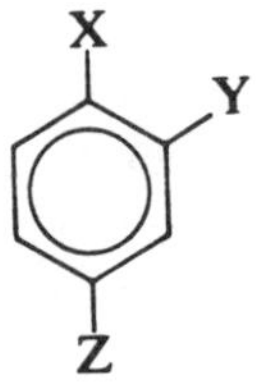

X and Y were electron donor groups and Z an electron accepting group, for instance, X = NH_2, OH, alkyl, halogen, alkoxy, Y = NH_2, OH, CH_3, Z = NO_2, CN, CF_3.

Derivatives of 4-nitroaniline (e.g. 3-methyl or 3-amino) are not sweet (Table 3.15).

2,4,6-Tribromo-3-carboxyalkylbenzamides and -3-carboxyalkoxy-benzamides were decribed [299] in 1983 as intensely sweet-tasting compounds in which the sweet potency depended on the nature of the substituents in position 3 (Table 3.16).

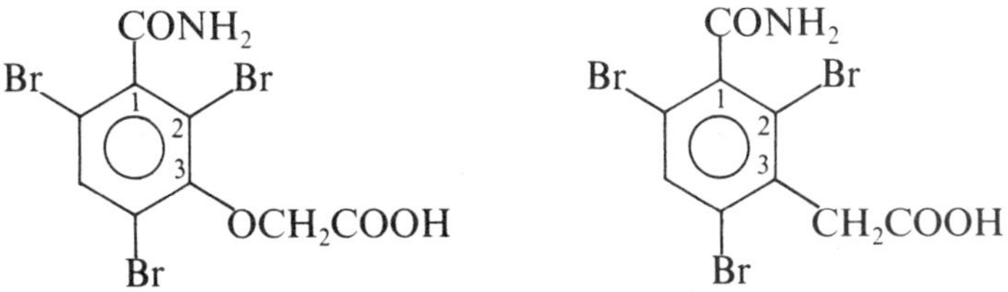

Table 3.16 — Relative sweetness of tribromobenzamides

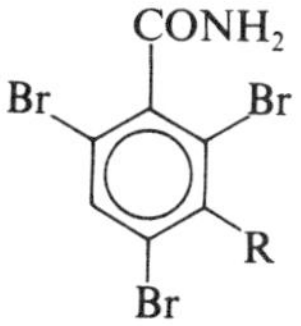

R	Relative sweetness[a]
CH_2COOH	100
CH_2CH_2COOH	4000
CH_2CH_2COONa	6000–7000[b]
$CH_2(CH_2)_2COOH$	400
$CH_2CH(CH_3)COOH$	1200
$CH_2CH(OC_2H_5)COOH$	200
$CH_2CH(OH)COOH$	1600
$CH_2CH(Br)COOH$	1600
$CH_2CH(NH_2)COOH$	40
$CH=CHCOOH$	4000
OCH_2COOH	400
$O(CH_2)_3COOH$	100

[a] Ref. [299]. [b] Ref. [300].

The synthesis starts with 3-nitrophenylacetic acid which is reduced and then brominated to give the corresponding amino derivative. Diazotization, followed by a Sandmeyer reaction yields the cyano derivative which is then hydrolysed, either by acid or alkali, to the 3-carbamoyl-2,4,6-tribromophenylacetic acid (Scheme 3.37).

3-Nitrophenylacetic acid can be prepared in 60 % yield from 3-nitroacetophenone under the conditions of a modified Willgerodt reaction [301], as shown in Scheme 3.38.

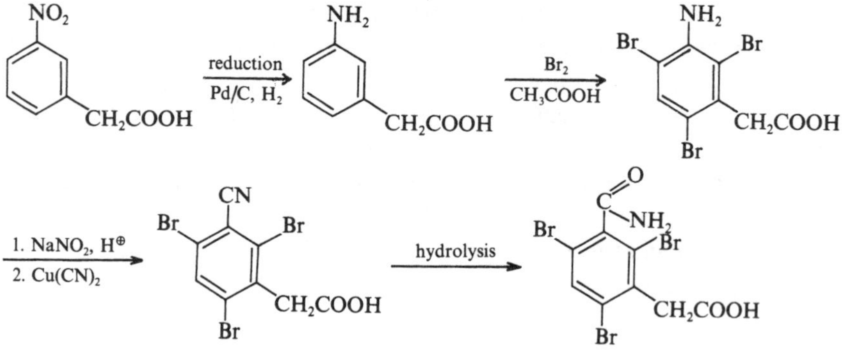

Scheme 3.37

Several other syntheses of trihalogen-substituted benzamides are published [300, 302].

Acute toxicity of such derivatives comes close to that of saccharin and cyclamates (LD_{50} in rats and cats 5–7 g/kg) [299]. The available data is too limited to allow any serious toxicological and hygienic assessment.

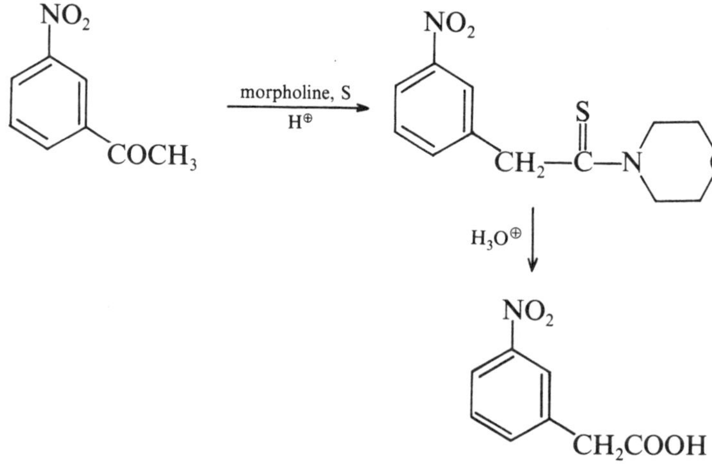

Scheme 3.38 m. p. 120–120° C

3.8 OTHER SYNTHETIC SWEET SUBSTANCES

A novel sweetener of the indole type, 3,4-dehydro-2-(3-indolylmethyl)-1-methylpiperidine-4-carboxylic acid, belongs to this class. Needless to say, its discovery, during the course of the synthesis of lysergic acid, was also fortuitous. Its sweet potency exceeds 500–1000 times that of sucrose [303].

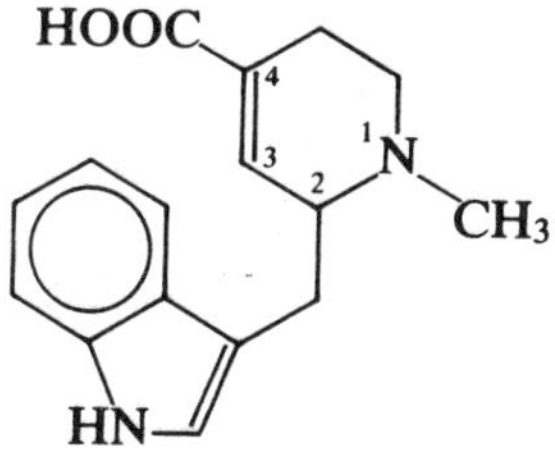

HOOC CH$_3$ HN

$M = 269$, m.p. = 215 °C

The preparation involves the condensation indole-3-carboxaldehyde with 3,4-dehydro-3-hydroxy-1-methylpiperidine-4-carboxylic acid ethyl ester, followed by a reduction (Pt/H_2), dehydration and the hydrolysis of the ethyl ester (Scheme 3.39).

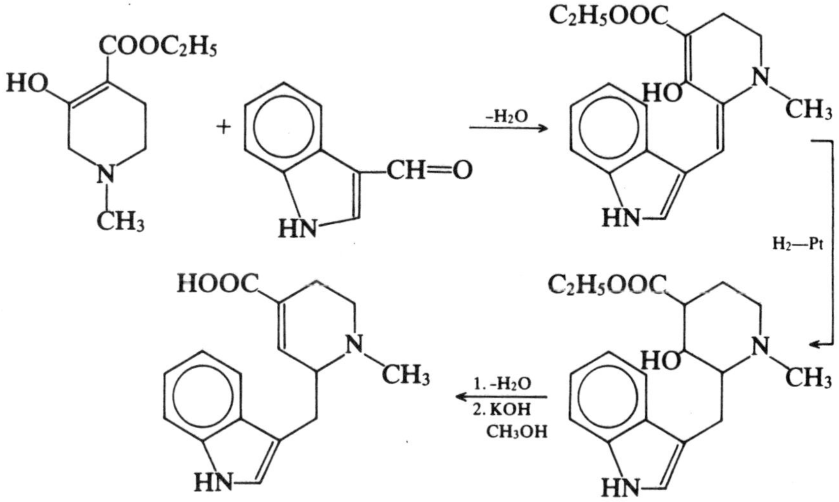

COOC$_2$H$_5$ HO N CH$_3$ + CH=O HN $\xrightarrow{-H_2O}$ C$_2$H$_5$OOC HO N CH$_3$ HN H$_2$—Pt C$_2$H$_5$OOC HO N CH$_3$ HN $\xrightarrow{1.\ -H_2O;\ 2.\ KOH,\ CH_3OH}$ HOOC N CH$_3$ HN

Scheme 3.39

The structure was confirmed by ^{1}H NMR and UV spectra. The preparation of the 3-hydroxy derivative was patented [23]; however, the biological tests are not yet complete.

The sodium salt of 2-(4-methoxybenzoyl)benzoic acid is 150 times sweeter than sucrose, water soluble and thermally stable;

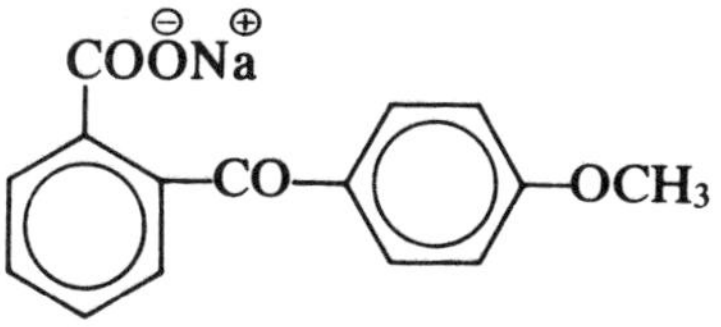

$COO^{\ominus}Na^{\oplus}$ —CO— —OCH$_3$

it belongs amongst the nitrogen-free sweeteners sometimes designated as S23/64 (Süßhilfe). The sweet taste changes to bitter at higher concentration (> 0.02 %) [286]. Although the salt supposedly poses no health hazards, no supporting biological tests have been published.

Glucine is a sodium salt of a mixture of the disulpho and trisulpho acids of triazine. Its yellow crystals melt at 223 °C and decompose at over 250 °C. Its sweetness is 300 times that of sucrose but it was banned for use because of its high toxicity [304].

3.8.1 Derivatives of Triazole and Tetrazole

Uteg and Gehlen [305] described in 1970 the preparation of a sweet compound in the 1,2,4-triazole series. The cyclization of 1-(2-hydroxybenzoyl)-5-ethoxycarbonyldiaminoguanidine yields 5-(2-hydroxyphenyl)-3-amino-4-ethoxycarbonylamino-1,2,4-triazole

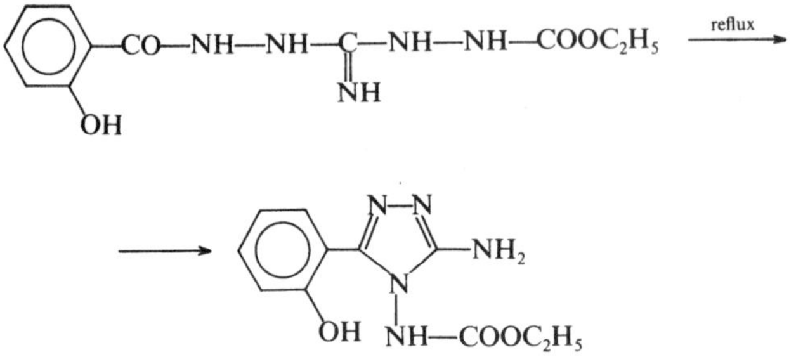

CO—NH—NH—C(=NH)—NH—NH—$COOC_2H_5$ (OH) —reflux→

NH$_2$, NH—$COOC_2H_5$, OH

m.p. 197–198° C

which is soluble in hot water and in hot ethanol and is 450–500 times sweeter than sucrose. It has a clean taste without any aftertaste. The data on possible applications or toxicity are not available.

5-Aminotetrazoles are also considered as potential sweeteners in the literature but yet again, no results of biological tests have been published [306].

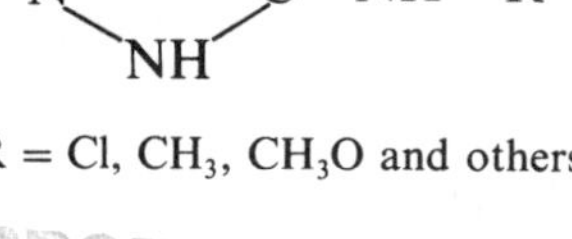

N—N, N, C—NH—R, NH

R = Cl, CH_3, CH_3O and others

REFERENCES

1. Rosival, L. & Szokolay, A. (1983) *Heterogeneous Compounds in Food.* Osveta, Martin (in Slovak).
2. Melichar, B. E. A. (1987) *Synthetic Drugs.* Avicenum, Prague (in Czech).
3. von Rymon Lipinski, G.-W. (1987) *Lebensmittelchem. Gerichtl. Chem.* **41**, 101.
4. Baines, C. J. (1984) *Can. Med. Assoc. J.* **131**, 1445.
5. Higginbotham, J. D. (1984) *Food Technol. Aust.* **36**, 552.
6. Newsome, R. L. (1986) *Food Technol.* (Chicago) 195.
7. Penttilä, P.-L., Salminen, S. & Niemi, E. (1988) *Z. Lebensm.-Unters. Forsch.* **186**, 11.
8. Sjöberg, A.-M. & Penttilä, P.-L. (1988) *Z. Lebensm.-Unters. Forsch.* **186**, 197.
9. Fahlberg, C. & Remsen, I. (1987) *Chem. Ber.* **12**, 469.
10. Remsen, I. & Fahlberg, C. (1979–1980) *J. Am. Chem. Soc.* **1**, 426.
11. Eur. Patent Appl. 81959 (1983) [*Chem. Abstr.* **99**, 1983, 157076].
12. Arnold, D. L., Krewski, D. & Munro, I. C. (1983) *Toxicology* **27**, 179.
13. Beck, K. M. (1983) In: Kirk, R. E. & Othmer, D. F. (eds) *Encyclopedia of Chemical Technology.* Vol. 22. J. Wiley & Sons Inc., New York, p. 448.
14. Fabiani, P. (1981) *Ann. Fals. Exp. Chim.* **74**, 273.
15. Radford, T., Cook, J. M., Dalsis, D. E., Wolf, E. & Voigt, E. (1985) *Food Chem. Toxicol.* **23**, 419.
16. Japan. Kokai 5959677 (1984) [*Chem. Abstr.* **101**, 1984, 90915].
17. Span. 509932 (1983) [*Chem. Abstr.* **99**, 1983, 86898].
18. Rose, N. C. & Rome, S. (1970) *J. Chem. Educ.* **47**, 649.
19. van der Wel, H., van der Heyden, A., & Peer H. G. (1987) *Food Rev. Int.* **3**, 193.
20. Rohse, H. & Belitz, H.-D. (1988) *Z. Lebensm.-Unters. Forsch.* **187**, 425.
21. Hamor, G. H. (1961) *Science* **134**, 1416.
22. Rossy, P. A., Hoffmann, W. & Mueller, N. (1980) *J. Org. Chem.* **45**, 617.
23. Crammer, B. & Ikan, R. (1977) *Chem. Soc. Rev.* **6**, 431.
24. Perez, C. F., Calandi, E. L., Mazzieri, M. R., Arguello, B., Suarez, A. R. & Fumarola, M. J. (1984) *Org. Prep. Proced. Int.* **16**, 37.
25. Kočna, A. (1987) *Čas. Lék. Čes.* **126**, 836 (in Czech).
26. Wannagat, U. (1985) *Nova Acta Leopold.* **59**, 353.
27. Nelson, J. J. (1976) *J. Assoc. Off. Anal. Chem.* **15**, 233.
28. Hettler, H., Anderson, D. W., Reynolds, W. A. & Filer, L. J. (1976) *Adv. Heterocycl. Chem.* **15**, 233.
29. Pitkin, R. M., Anderson, D. W., Reynolds, W. A. & Filer, L. J. (1971) *Proc. Soc. Exp. Biol. Med.* **137**, 803.
30. Chay, M. & Hayert, L. (1988) *Bios* (Nancy) **19**, 29.
31. Redlinger, P. A. & Setser, C. S. (1987) *J. Food Sci.* **52**, 1391.
32. Bart, J. C. J. (1968) *J. Chem. Soc.,* B 376.
33. Okaya, Y. R. (1969) *Acta Crystallogr.,* B **25**, 2257.
34. Rader, C. R., Tihanyi, S. G. & Zienty, F. B. (1967) *J. Food Sci.* **32**, 357.
35. Alary, J., Arnaud, J. & Villet, A. (1981) *Bull. Trav. Soc. Pharm.* (Lyon) **25**, 58, 65.
36. Anon. (1984) *Z. Lebensm.-Unters. Forsch.* **179**, 453.
37. Stolzberg, R. J. (1986) *J. Chem. Educ.* **63**, 351.
38. Görner, F. (1987) *Bull. Food Res.* (Bratislava) **26**, 251.
39. Smyly, D. S., Woodward, B. B. & Conrad, E. C. (1976) *J. Assoc. Off. Anal. Chem.* **59**, 14.
40. Woodward, B. B., Heffelfinger, G. P. & Ruggles, D. I. (1979) *J. Assoc. Off. Anal. Chem.* **62**, 1011.

41. Örsi, F., Ember-Karpati, M., Lasztity, R. & Abraham-Szabo, A. (1983) *Elelm. Ip.* **37**, 41 (in Hungarian).
42. Delaney, M. F., Pasko, K. M., Mauro, D. M., Gsell, D. S., Korologos, P. C., Morawski, J. & Krolikowski, L. J. (1985) *J. Chem. Educ.* **62**, 618.
43. Hann, J. T. & Gilkison, I. S. (1987) *J. Chromatogr.* **395**, 317.
44. Terada, H. & Sakabe, Y. (1985) *J. Chromatogr. Anal. Chem.* **346**, 333.
45. Holak, W. & Krinitz, B. (1980) *J. Assoc. Off. Anal. Chem.* **63**, 163.
46. Rubach, K. & Offizorz, P. (1983) *Dtsch. Lebensm.-Rdsch.* **79**, 88.
47. Zache, U. & Gründing, H. (1987) *Z. Lebensm.-Unters. Forsch.* **184**, 503.
48. Rao, M. V., Krishnamacharyulu, A. G., Kapur, O. P. & Sastry, C. S. P. (1983) *J. Food Sci. Technol.* **20**, 202.
49. Suzuki, H. & Suzuki, N. (1988) *Mutat. Res.* **209**, 13.
50. Cohen, S. M., Arai, M., Jacob, J. B. & Friedell, G. H. (1979) *Cancer Res.* **39**, 1207.
51. Hagiwara, A., Fukushima, S., Kitaori, M., Shibata, M. & Ito, N. (1984) *Gann* **75**, 763.
52. Šula, J. (1982) *Chem. Listy* **76**, 962 (in Czech).
53. Hicks, R. M., Chovanie, J. & Wakefield, J. (1978) *Mechanism of Tumor Promotion and Cocarcinogenesis* **2**, 457.
54. Audrieth, L. F. & Sveda, M. (1944) *J. Org. Chem.* **9**, 89.
55. Beck, K. M. (1969) In: Kirk, R. E. & Othmer, D. F. (eds) *Encyclopedia of Chemical Technology*. J. Wiley & Sons Inc., New York.
56. Myška, J., Jokl, J. & Kalina, O. (1967) *Chem. Listy* **61**, 1489 (in Czech).
57. Ger. Offen. 3403813 (1986) [*Chem. Abstr.* **104**, 1986, 87377].
58. Nofre, C. & Pautet, F. (1975) *Bull. Soc. Chim. Fr.* 686.
59. Nofre, C. & Pautet, F. (1975) *Naturwissenschaften* **62**, 97.
60. Pautet, F. & Nofre, C. (1978) *Z. Lebensm.-Unters. Forsch.* **166**, 167.
61. Pautet, F. & Nofre, C. (1978) *Pharm. Acta Helv.* **53**, 231.
62. Unterhalt, B. & Böschemeyer, L. (1972) *Naturwissenschaften* **59**, 271.
63. Unterhalt, B. & Böschemeyer, L. (1972) *Z. Lebensm.-Unters. Forsch.* **149**, 227.
64. Unterhalt, B. & Böschemeyer, L. (1976) *Z. Lebensm.-Unters. Forsch.* **161**, 275.
65. U.S. 2800501 (1957) [*Chem. Abstr.* **51**, 1957, 17987].
66. U.S. 3082247 (1963) [*Chem. Abstr.* **59**, 1963, 8616].
67. Benson, G. A. & Spillane, W. J. (1976) *J. Med. Chem.* **19**, 869.
68. de Nardo, M., Runti, C. & Ulian, F. (1984) *Farmaco Ed. Sci.* **39**, 125.
69. Pautet, F. & Dandon, M. (1985) *Pharmazie* **40**, 428.
70. Pautet, F. & Dandon, M. (1986) *Pharm. Acta Helv.* **61**, 215.
71. Urban, J. & Plánička, K. (1967) *Českoslov. Farm.* **16**, 48 (in Czech).
72. Richardson, M. L. (1967) *Talanta* **14**, 385.
73. Hoo, D. & Hu, C. (1972) *Anal. Chem.* **44**, 2111.
74. Greve, H. & Bonifer, P. (1987) *Flüssiges Obst* **54**, 382.
75. von Rymon Lipinski, G.-W. & Brixius, H.-C. (1979) *Z. Lebensm.-Unters. Forsch.* **168**, 212.
76. Ferrando, R. (1981) *Ann. Fals. Exp. Chim.* **74**, 285.
77. Herrmann, A., Damawandi, E. & Wagmann, M. (1983) *J. Chromatogr.* **280**, 85.
78. Biemer, T. A. (1989) *J. Chromatogr.* **463**, 463.
79. Renwick, A. G. (1985) *Food Chem.* **16**, 281.
80. Clauss, K. & Jensen, H. (1973) *Angew. Chem.* **85**, 965.
81. Kobelt, D., Paulus, E. F. & Clauss, K. (1971) *Tetrahedron Lett.* 3627.
82. Clauss, K., Lueck, E. & von Rymon Lipinski, G. W. (1976) *Z. Lebensm.-Unters. Forsch.* **162**, 37.

83. Arpe, H. J. (1978) In: B. Guggenheim (ed.) *Proc. Health Sug. Subst., ERGOB Conf.*, Geneva. Karger, Basel, p. 178.
84. von Rymon Lipinski, G.-W. (1984) *Spec. Chem.* 39 [*Chem. Abstr.* **102**, 1985, 130501].
85. Clauss, K. (1980) *Ann. Chem.* 494.
86. von Rymon Lipinski, G.-W. (1985) *Food Chem.* **16**, 259.
87. von Rymon Lipinski, G.-W. (1986) *Alimenta* **25**, 150.
88. Suhaj, M., Kováč, M., Uher, M. & Vácová, T. (1986) *Bull. Food Res. (Special issue)* (Bratislava) 13.
89. Suhaj, M. & Kováč, M. (1988) *Bull. Food Res.* (Bratislava) **27**, 31 (in Slovak).
90. Suhaj, M., Bubelíniová, E. & Kováč, M. (1988) *Bull. Food Res.* (Bratislava) **27**, 37 (in Slovak).
91. Hoppe, K. & Gassmann, B. (1985) *Nahrung* **29**, 417.
92. Fed. Regist. 28 July 1988 5328379 [*Chem. Abstr.* **109**, 1988, 148108].
93. Bubelíniová, E., Suhaj, M. & Kováč, M. (1988) *Bull. Food Res.* (Bratislava) **27**, 23 (in Slovak).
94. Ger. Offen. 3531358 (1986) [*Chem. Abstr.* **105**, 1986, 24281].
95. Ger. Offen. 3531358 (1987) [*Chem. Abstr.* **107**, 1987, 134331].
96. Grosspietsch, H. & Hachenberg, H. (1980) *Z. Lebensm.-Unters. Forsch.* **171**, 41.
97. Paulus, E. F. (1975) *Acta Crystallogr.*, B **31**, 1191.
98. Spillane, W. J. & McGlinchey, G. (1981) *J. Pharm. Sci.* **70**, 933.
99. Inglett, G. E. (1981) *Food Technol.* (London) 37.
100. Hoppe, K. & Gassmann, B. (1985) *Lebensmittelindustrie* **32**, 227.
101. Joachim, J., Kalantzis, G., Joachim, G., Delonca, H., Maury, L. & Rambaud, J. (1987) *Pharm. Acta Helv.* **62**, 262.
102. Mazur, R. H. & Ripper, A. (1979) *Dev. Sweeteners* **1**, 87.
103. Mazur, R. H., Schlatter, J. M. & Goldkamp, A. H. (1969) *J. Am. Chem. Soc.* **91**, 2684.
104. Mazur, R. H. (1984) In: Stegink, L. J. & Filler, L. J., Jr. (eds) *Aspartame. Physiology and Biochemistry*. Marcel Dekker, New York.
105. Davey, J. M., Laird, A. H. & Morley, J. S. (1966) *J. Chem. Soc., C* 555.
106. Franke, R. (1978) *Chem. Listy* **72**, 945 (in Czech).
107. Davídková, E. & Prudel, M. (1978) *Prům. Potr.* **29**, 367 (in Czech).
108. Knopp, R. H., Brandt, K. & Arky, R. A. (1976) *J. Toxicol. Environ. Health* **2**, 417.
109. U.S. 3753739 (1973) [*Chem. Abstr.* **80**, 1974, 13818].
110. Ger. 2264394 (1973) [*Chem. Abstr.* **80**, 1974, 58607].
111. Vetsch, W. (1985) *Food Chem.* **16**, 245.
112. Davídková, E., Prudel, M. & Volfová, Z. (1984) *Sbornik ÚVTIZ, Potrav. Vědy* **2**, 49.
113. Swed. 508590 (1971) [*Ref. Zh. Khim.* 1972, 2H125].
114. Neth. 70071 (1971) [*Chem. Abstr.* **76**, 1972, 86150].
115. Brit. 1339101 (1973) [*Chem. Abstr.* **80**, 1974, 96372].
116. Ger. 2256055 (1973) [*Chem. Abstr.* **79**, 1973, 42841].
117. Ger. 2452285 (1975) [*Chem. Abstr.* **83**, 1975, 79615].
118. Eur. Pat. Appl. 130464 (1985) [*Chem. Abstr.* **103**, 1985, 37737].
119. Belg. 898650 (1984) [*Chem. Abstr.* **101**, 1984, 37737].
120. Ariyoshi, Y., Yamatani, T., Uchiyama, N., Adachi, Y. & Sato, N. (1973) *Bull. Chem. Soc. Jpn.* **46**, 1893.
121. Ariyoshi, Y., Yamatani, T. & Adachi, Y. (1973) *Bull. Chem. Soc. Jpn.* **46**, 2611.
122. U.S. 3901871 (1975) [*Ref. Zh. Khim.* 1976, 11021].

123. Ger. 2233535 (1973) [*Chem. Abstr.* **78**, 1973, 98023].
124. Brit. 1481186 (1976) [*Ref. Zh. Khim.* 1978, 8013].
125. U.S. 3809190 (1974).
126. Ger. 2326897 (1972) [*Chem. Abstr.* **80**, 1974, 60216].
127. U.S. 3833554 (1974) [*Ref. Zh. Khim.* 1975, 14024].
128. Ger. 2152111 (1970) [*Chem. Abstr.* **77**, 1972, 20029].
129. Ariyoshi, Y. & Sato, N. (1972) *Bull. Chem. Soc. Jpn.* **45**, 942.
130. Ariyoshi, Y. & Sato, N. (1972) *Bull. Chem. Soc. Jpn.* **45**, 2015.
131. Japan 25190 (1973) [*Chem. Abstr.* **80**, 1974, 3798].
132. Furda, I., Malizia, P. D., Kolov, M. G. & Vernieri, P. (1975) *J. Agric. Food Chem.* **23**, 340.
133. Brit. 1298700 (1972) [*Chem. Abstr.* **78**, 1973, 111764].
134. Japan 00812 (1973) [*Chem. Abstr.* **78**, 1973, 148238].
135. Eur. Pat. Appl. 92933 (1983) [*Chem. Abstr.* **100**, 1984, 103908].
136. Ger. Offen. 3600731 (1987) [*Chem. Abstr.* **107**, 1987, 176486].
137. Japan 7396557 (1973) [*Chem. Abstr.* **80**, 1974, 96371].
138. Eur. Pat. Appl. 196866 (1986) [*Chem. Abstr.* **107**, 1987, 7612].
139. Ger. Offen. 3545193 (1987) [*Chem. Abstr.* **107**, 1987, 176485].
140. Japan 42491 (1974) [*Chem. Abstr.* **83**, 1975, 10848].
141. Pietsch, H. (1976) *Tetrahedron Lett.* 4053.
142. Vinick, F. J. & Jung, S. (1982) *Tetrahedron Lett.* **23**, 1315.
143. U.S. 3475403 (1969) [*Ref. Zh. Khim.* 1970, 23H323].
144. U.S. 3492131 (1970) [*Ref. Zh. Khim.* 1971, 4H348].
145. Fr. 2114657 (1972) [*Chem. Abstr.* **78**, 1973, 84821].
146. Pavlova, L. A., Komarova, T. V., Davidovich, Yu. A. & Rogozhin, S. B. (1981) *Usp. Khim.* **4**, 590.
147. U.S. 4021418 (1976) [*Chem. Abstr.* **85**, 1976, 124370].
148. Japan 6163694 (1986) [*Chem. Abstr.* **105**, 1986, 227315].
149. Japan 6274296 (1987) [*Chem. Abstr.* **107**, 1987, 95611].
150. Japan 6274297 (1987) [*Chem. Abstr.* **107**, 1987, 114543].
151. Span. 550895 (1988) [*Chem. Abstr.* **108**, 1988, 38435].
152. Japan 63145298 (1988) [*Chem. Abstr.* **109**, 1988, 211484].
153. Japan 62164692 (1988) [*Chem. Abstr.* **108**, 1988, 6436].
154. Chen Shui Tein, Li Kuo Fang & Wang Kung Tsung (1988) *J. Chin. Chem. Soc.* (Taipei) **35**, 207 [*Chem. Abstr.* **110**, 1989, 20744].
155. Isowa, Y., Ohmori, M., Ichikawa, T., Mori, K., Nonaka, Y., Kihara, K., Oyama, K., Satoh, H. & Nishimura, S. (1979) *Tetrahedron Lett.* 2611.
156. Oyama, K., Kihara, K. & Nonaka, Y. (1981) *J. Chem. Soc., Perkin Trans. 2* 356.
157. Oyama, K., Nishimura, S., Nonaka, Y., Kihara, K. & Hashimoto, T. (1981) *J. Org. Chem.* **46**, 5241.
158. Nakanishi, K., Kamikubo, T. & Matsuno, R. (1985) *Biotechnology* **3**, 459.
159. Mapelli, C., Newton, M. G., Ringold, C. E. & Stammer, G. H. (1987) *Int. J. Pept. Protein Res.* **30**, 498.
160. King, S. W. & Stammer, C. H. (1981) *J. Org. Chem.* **46**, 4780.
161. Schiffmann, S. S. & Englehard, H. H. (1976) *Physiol. Behav.* **17**, 523.
162. Brit. 1229265 (1972) [*Chem. Abstr.* **78**, 1973, 111760].
163. Fujino, M., Wakimasu, M., Tanaka, K., Aoki, H. & Nakajima, N. (1973) *Naturwissenschaften* **60**, 351.
164. Neth. Appl. 7012899 (1972) [*Chem. Abstr.* **77**, 1972, 33111].
165. Mazur, R. M., Goldkamp, A. H., James, P. A. & Schlatter, J. M. (1970) *J. Med.*

Chem. **13**, 1217.
166. Mazur, R. H., Reuter, J. A., Swiatek, K. A. & Schlatter, J. M. (1973) *J. Med. Chem.* **16**, 1284.
167. Ariyoshi, Y., Yasuda, N. & Yamatani, T. (1974) *Bull. Chem. Soc. Jpn.* **47**, 326.
168. Eur. Pat. Appl. 34876 (1982) [*Chem. Abstr.* **96**, 1982, 104780].
169. Miyoshi, M., Nunami, K., Sugano, H. & Fujii, T. (1978) *Bull. Chem. Soc. Jpn.* **51**, 1433.
170. Fujino, M., Wakimasu, M., Mano, M., Tanaka, K., Nakajima, N. & Aoki, H. (1976) *Chem. Pharm. Bull.* **24**, 2112.
171. MacDonald, S. A., Willson, C. G., Chorev, M., Vernacchia, F. S. & Goodman, M. (1980) *J. Med. Chem.* **23**, 413.
172. van der Heijden, A., Brussel, L. B. P. & Peer, H. G. (1979) *Chem. Senses Flavour* **4**, 141.
173. Iwamura, H. (1981) *J. Med. Chem.* **24**, 572.
174. Lelj, F., Tancredi, T., Temussi, P. A. & Toniolo, C. (1976) *J. Am. Chem. Soc.* **98**, 6669.
175. Ariyoshi, Y. (1976) *Agric. Biol. Chem.* **40**, 983.
176. Ariyoshi, Y. (1980) *Agric. Biol. Chem.* **44**, 943.
177. Wieser, H., Jugel, H. & Belitz, H.-D. (1977) *Z. Lebensm.-Unters. Forsch.* **164**, 277.
178. Ariyoshi, Y. (1984) *Bull. Chem. Soc. Jpn.* **57**, 3197.
179. Ariyoshi, Y. (1985) *Bull. Chem. Soc. Jpn.* **58**, 1727.
180. Ariyoshi, Y. (1986) *Bull. Chem. Soc. Jpn.* **59**, 1027.
181. Siemion, I. Z. & Picur, B. (1984) *Pol. J. Chem.* **58**, 475.
182. Kawai, M., Chorew, M., Marin-Rose, J. & Goodman, M. (1980) *J. Med. Chem.* **23**, 420.
183. Thomsen, M. W., Dalton, J. M. & Stewart, C. N. (1988) *Chem. Senses* **13**, 397.
184. Ger. Offen. 1948788 (1970) [*Chem. Abstr.* **75**, 1971, 19014].
185. Neth. 700716 (1971) [*Chem. Abstr.* **76**, 1972, 86150].
186. Can. 948185 (1974) [*Chem. Abstr.* **82**, 1975, 86641].
187. Ger. 1936159 (1970) [*Chem. Abstr.* **72**, 1970, 101098].
188. U.S. 379918 (1974) [*Ref. Zh. Khim.* 1975, 2H84].
189. Tsang, J. W., Schmied, B., Nyfeler, R. & Goodman, M. (1984) *J. Med. Chem.* **27**, 1663.
190. Chorew, M., Wilson, C. G. & Goodman, M. (1977) In: Goodman, M. & Meienhofer, J. (eds) *Peptides. Proc. 5th Am. Pept. Symp.* J. Wiley & Sons Inc., New York, p. 5.
191. Brussel, L. B. P., Peer, H. G. & van der Heijden, A. (1975) *Z. Lebensm.-Unters. Forsch.* **159**, 337.
192. van der Heijden, A., Brussel, L. B. P. & Peer, H. G. (1978) *Food Chem.* **3**, 207.
193. van der Heijden, A., van der Wel, H. & Peer, H. G. (1985) *Chem. Senses* **10**, 57.
194. Miyashita, Y., Takahashi, Y., Takayama, C., Sumi, K., Nakatsuka, K., Ohkubo, T., Abe, H. & Sasaki, S. (1986) *J. Med. Chem.* **29**, 906.
195. Shinoda, I. & Okai, H. (1984) In: Munakata, E. (ed.) *Peptide Chemistry*. Protein Research Foundation, Osaka, p. 135.
196. U.S. 3798204 (1974) [*Ref. Zh. Khim.* 1975, 3P267].
197. U.S. 3853835 (1974) [*Chem. Abstr.* **81**, 1974, 58606].
198. Mioyshi, M., Numani, K., Sugano, H. & Fujii, T. (1978) *Bull. Chem. Soc. Jpn.* **51**, 1433.
199. Sukehiro, M., Minematsu, H. & Noda, K. (1977) *Seikatsu Kagaku* **11**, 9 [*Chem. Abstr.* **87**, 1977, 168407].

200. Loudon, G. M., Radhakrishna, A. S., Almond, M. R., Blodgett, J. K. & Boutin, R. H. (1984) *J. Org. Chem.* **49**, 4272, 4277.
201. Fuller, W. D., Goodman, M. & Verlander, M. S. (1985) *J. Am. Chem. Soc.* **107**, 5821.
202. Goodman, M., Bland, J., Tsang, J., Goddington, J., Temussi, A. P., Tancredi, T., Lelj, F., Fuller, W. D. & Verlander, M. S. (1985) In: Deber, C. M., Hruby, V. J. & Kopple, K. D. (eds) *Peptides: Structure and Function. Proc. 9th Am. Pept. Symp.* Pierce Chem. Co., Rockford, p. 725.
203. Rizzi, G. P. (1985) *J. Agric. Food Chem.* **33**, 19.
204. Rizzi, G. P. & Echler, R. S. (1986) *Food Chem.* **20**, 165.
205. Lapidus, M. & Sweeney, J. M. (1973) *J. Med. Chem.* **16**, 163.
206. de Nardo, M., Runti, C., Ulian, F. & Vio, L. (1976) *Farmaco* (Pavia) **31**, 906.
207. Tinti, J. M., Durozard, D. & Nofre, C. (1980) *Naturwissenschaften* **67**, 193.
208. Tinti, J. M., Durozard, D. & Nofre, C. (1981) *Naturwissenschaften* **68**, 143.
209. Kawai, M., Nyfeler, R., Berman, J. M. & Goodman, M. (1982) *J. Med. Chem.* **25**, 397.
210. Rodriguez, M. & Goodman, M. (1984) *J. Med. Chem.* **25**, 1668.
211. Goodman, M., Rodriguez, M. & Tsang, J. (1984) *Peptides* 549.
212. Ciajolo, M. R., Lejl, F., Tancredi, T., Temussi, P. A. & Tuzi, A. J. (1983) *J. Med. Chem.* **26**, 1060.
213. Ger. 2355010 (1974) [*Chem. Abstr.* **80**, 1974, 121331].
214. U.S. 3907766 (1975) [*Ref. Zh. Khim.* 1976, 12P316].
215. U.S. 3801563 (1974) [*Ref. Zh. Khim.* 1975, 3011].
216. U.S. 3959245 (1976) [*Ref. Zh. Khim.* 1977, 3H84].
217. Fujino, M., Wakimasu, M., Tanaka, K., Aoki, H. & Nakajima, N. (1973) In: Kotake, H. (ed.) *Proc. XIth Symposium on Peptide Chemistry*, Tokyo, p. 103.
218. Bláha, K., Glanzová, J. & Pospíšek, J. (1979) *Chem. Listy* **73**, 701 (in Czech).
219. Japan 61200999 (1985) [*Chem. Abstr.* **106**, 1987, 156864].
220. Japan 61291596 (1985) [*Chem. Abstr.* **106**, 1987, 176872].
221. Eur. Pat. Appl. 256475 (1986) [*Chem. Abstr.* **109**, 1988, 211487].
222. Japan 6339896 (1988) [*Chem. Abstr.* **109**, 1988, 170930].
223. Hatada, M., Jancarik, J., Graves, B. & Kim, S. H. (1985) *J. Am. Chem. Soc.* **107**, 4279.
224. Boehm, M. F. & Bada, J. L. (1984) *Proc. Natl. Acad. Sci. U.S.A.* **81**, 5263.
225. Gaines, S. M. & Bada, J. L. (1988) *J. Org. Chem.* **53**, 2757.
226. Frey, G. H. (1976) *J. Toxicol. Environ. Health* 401.
227. Daabees, T. T., Finkelstein, M. W., Stegink, L. D. & Applebaum, A. E. (1985) *Food Chem. Toxicol.* **23**, 887.
228. Stegink, L. D. (1987) *Food Technol.* (Chicago) **41**, 120.
229. Fernstrom, J. D., Fernstrom, M. H. & Gillis, M. A. (1983) *Life Sci.* **32**, 165.
230. Yokogoshi, H., Roberts, C. H., Caballero, B. & Wurtman, R. J. (1984) *Am. J. Clin. Nutr.* **40**, 1.
231. Stegink, L. D., Filler, L. J., Jr. & Baker, G. L. (1977) *J. Nutr.* **107**, 1837.
232. Filler, L. J., Jr., Baker, G. L. & Stegink, L. D. (1983) *J. Nutr.* **113**, 1591.
233. Brunner, R. L., Vorhees, C. V., Kinney, L. & Butcher, R. E. (1979) *Neurobehav. Toxicol.* **1**, 79.
234. Potts, W. J., Bloss, J. L. & Nutting, E. F. (1980) *J. Environ. Pathol. Toxicol.* **3**, 341.
235. Ishii, T. K., Usami, S. & Fujimoto, T. (1981) *Toxicology* **21**, 91.
236. Wurtman, R. J. (1986) *Food Chem. News* 65.
237. Coulombe, R. A. & Sharma, R. P. (1986) *Toxicol. Appl. Pharmacol.* **83**, 79.

238. Homler, B. E., see ref. 104, 254.
239. Güven, K. C. & Özol, T. (1984) *Acta Pharm. Turc.* **26**, 28.
240. Özol, T. (1984) *Acta Pharm. Turc.* **26**, 59.
241. Vachek, J. (1984) *Českoslov. Farm.* **33**, 217 (in Czech).
242. Počtová, M. & Kakáč, V. (1982) *Českoslov. Farm.* **31**, 113 (in Czech).
243. Nishijima, M., Kanmuri, M., Takamashi, S., Kaminura, H., Nakazato, M., Watari, Y. & Kimura, Y. (1976) *Shokuhin Eiseigaku Zasshi* **17**, 78 [*Chem. Abstr.* **85**, 1976, 19174].
244. Prudel, M. (1984) Stability and Applications of Sweeteners in Food. *PhD Thesis.* Organic Institute of Chemical Technology, Prague (in Czech).
245. Sherma, J., Chapin, S. & Follweiler, J. M. (1985) *Am. Lab.* **17**, 131 [*Chem. Abstr.* **102**, 1985, 165329].
246. Ciranni-Signoretti, E., Dell'Utri, A. & De Salvo, A. (1983) *Bull. Chim. Farm.* **122**, 289.
247. Škarka, P., Šestáková, I. & Turza, P. (1979) *Biol. Chem. Vet.* **XV (XXI)** 39 (in Czech).
248. Ibe, A., Saito, K., Nakazato, M., Kikuchi, Y., Fujinuma, K., Naoi, Y. & Nishima, T. (1985) *Shokuhin Eiseigaku Zasshi* **26**, 1.
249. Daniels, D. M., Joe, F. L., Jr., Warner, C. R. & Fazio, T. (1984) *J. Assoc. Off. Anal. Chem.* **67**, 513.
250. Fletcher, P. L. & Moore, S. (1975) In: Walter, R. & Meienhofer, J. (eds) *Peptides: Chemistry, Structure and Biology. Proc. 4th Am. Pept. Symp.* Ann Arbor Sci. Publ., Ann Arbor, p. 625.
251. Ishiwata, A. & Suzuki, Y. (1975) *Shokuhin Eiseigaku Zasshi* **16**, 420.
252. Veselý, Z., Davídková, E. & Prudel, M. (1980) *Nahrung* **24**, 525.
253. Škarka, P., Šestáková, I., Smolek, P. & Entlicher, G. (1980) *Biol. Chem. Vet.* **XVI (XXII)** 45 (in Czech).
254. Scherz, J. C., Monti, J. C. & Jost, R. (1983) *Z. Lebensm.-Unters. Forsch.* **177**, 124.
255. Fox, L., Anthony, G. D. & Lau, E. P. K. (1978) *J. Assoc. Off. Anal. Chem.* **59**, 1048.
256. Verzella, G. & Mangia, A. (1985) *J. Chromatogr.* **346**, 417.
257. Verzella, G., Bagnasco, G. & Mangia, A. (1985) *J. Chromatogr.* **349**, 83.
258. Prudel, M. & Davídková, E. (1985) *Nahrung* **29**, 381.
259. Cross, R. & Cunico, B. (1984) *LC, Liq. Chromatogr. HPLC Mag.* **2**, 678 [*Chem. Abstr.* **102**, 1984, 22915].
260. Terada, H. & Sakabe, Y. (1983) *Eisei Kohagu* **29**, 394.
261. Webb, N. G. & Beckman, D. D. (1984) *J. Assoc. Off. Anal. Chem.* **67**, 510.
262. Issag, H. J., Weiss, D., Ridlon, C., Fox, S. D. & Muschik, G. M. (1986) *J. Liq. Chromatogr.* **9**, 1791.
263. Tamase, K., Kitada, Y., Sasaki, M., Ueda, Y. & Takeshita, R. (1985) *Shokuhin Eiseigaku Zasshi* **26**, 515.
264. Herrmann, A., Damawandi, E. & Wagmann, M. (1983) *J. Chromatogr.* **280**, 85.
265. Argoudelis, C. J. (1984) *J. Chromatogr.* **303**, 256.
266. Tyler, A. T. (1984) *J. Assoc. Off. Anal. Chem.* **67**, 745.
267. Delaney, M. F., Pasko, K. M., Mauro, D. M., Gsell, D. S., Korogolos, P. C., Morawski, J., Krolikowski, L. J. & Warren, F. V., Jr. (1985) *J. Chem. Educ.* **62**, 618.
268. Tsang, W. S., Clarke, M. A. & Parrish, F. W. (1985) *J. Agric. Food Chem.* **33**, 734.
269. Lawrence, J. M. & Charbonneau, C. F. (1988) *J. Assoc. Off. Anal. Chem.* **71**, 934.
270. Baďura, J. (1975) Microbial Degradation of Methyl-Aspartyl-L-Phenylalanine. *PhD Thesis.* Institute of Chemical Technology, Prague (in Czech).
271. Boček, P. (1981) In: Everaerts, F. M. (ed.) *Anal. Chem. Symp. Ser.*, Vol. 6. Elsevier,

Amsterdam, p. 62.
272. Kvasnička, F. (1987) *J. Chromatogr.* **390**, 237.
273. Renneberg, R., Ridel, K. & Scheller, F. (1985) *Appl. Microbiol. Biotechnol.* **21**, 180.
274. Guibault, G. G., Lubrano, G. J., Kauffmann, J.-M. & Patriarche, G. J. (1988) *Anal. Chim. Acta* **206**, 369.
275. Berlinerblau, J. (1884) *J. Prakt. Chem.* **30**, 103.
276. Goldsmith, R. H. (1987) *J. Chem. Educ.* **64**, 954.
277. Klages, A. (1948) *Pharm. Zentralhalle Dtsch.* **87**, 161.
278. Rao, M. V., Kapur, O. P. & Prakasa Sastry, C. S. (1984) *J. Food Sci. Technol.* **21**, 148.
279. Krishnamurthy, S. (1982) *J. Chem. Educ.* **59**, 543.
280. Tapodó, J. (1977) *Édesipar* **28**, 2.
281. Marquez, G., Antonia, M., Guillen, S. R. & Guzman, C. M. (1986) *Afinidad* **43**, 63 [*Chem. Abstr.* **104**, 1986, 167004].
282. Noyce, W. K., Coleman, C. H. & Barr, J. T. (1951) *J. Am. Chem. Soc.* **73**, 1295.
283. Petersen, S. & Mueller, E. (1948) *Chem. Ber.* **81**, 31.
284. Möhler, K. (1950) *Z. Lebensm.-Unters. Forsch.* **90**, 431.
285. Tinti, J. M., Nofre, C. & Peytavi, A. M. (1982) *Z. Lebensm.-Unters. Forsch.* **175**, 266.
286. Hrdlička, J. (1973) *Výž. Lidu* **28**, 9 (in Czech).
287. Hengartner, U., Batcho, A. D., Blount, J. F., Leimgruber, W., Larscheid, M. E. & Scott, J. W. (1979) *J. Org. Chem.* **44**, 3748.
288. Finley, J. W. & Friedman, M. (1973) *Agric. Food Chem.* **21**, 33.
289. Dalgliesh, C. E. (1952) *J. Chem. Soc.* 137.
290. Bragg, R. W., Chow, Y., Dennis, L., Ferguson, L. N., Morga, G., Ogino, C., Pugh, H. & Winters, M. (1978) *J. Chem. Educ.* **55**, 281.
291. Blanksma, J. J. & van der Weyden, P. W. M. (1940) *Rec. Trav. Chim.* **59**, 629.
292. Verkade, P. E., van Dijk, C. P. & Meerburg, W. (1946) *Rec. Trav. Chim.* **65**, 346.
293. Lawrence, A. R. & Ferguson, L. N. (1960) *J. Org. Chem.* **25**, 1220.
294. Blanksma, J. J. (1946) *Rec. Trav. Chim.* **65**, 203.
295. Blanksma, J. J. & Hoegen, D. (1946) *Rec. Trav. Chim.* **65**, 333.
296. Blanksma, J. J. & Petri, E. M. (1947) *Rev. Trav. Chim.* **66**, 365.
297. U.S. 4633006 (1986) [*Chem. Abstr.* **106**, 1987, 196038].
298. Eur. Pat. Appl. 146913 (1985) [*Chem. Abstr.* **104**, 1986, 19509].
299. Gries, H., Mützel, W., Belitz, H. D., Wieser, H., Krauss, I. & Stemfl, W. (1983) *Z. Lebensm.-Unters. Forsch.* **176**, 376.
300. Eur. Pat. Appl. 55689 (1982) [*Chem. Abstr.* **98**, 1983, 88993].
301. Singh, B. (1978) *Synth. Commun.* **8**, 275.
302. Belitz, H. D., Rohse, H., Stempfl, W. & Wieser, H. (1986) *Getreide, Mehl, Brot* **40**, 371.
303. Hofmann, A. (1972) *Helv. Chim. Acta* **55**, 2934.
304. Lenges, J. (1986) *Chim. Nouv.* **4**, 483.
305. Uteg, K. H. & Gehlen, H. (1970) *Arch. Pharm.* **303**, 634.
306. Eur. Pat. Appl. 299533 (1988) [*Chem. Abstr.* **110**, 1989, 212829].

4

Multiple Sweeteners

Sucrose, the 'royal' carbohydrate, about which Disraeli said that it "charms infancy and soothes old age" is the standard of sweetness for us. To all but chemists, sucrose is sugar and its clean, sweet and uncomplicated taste is generally considered 'natural', whereas all other sweeteners taste artificially to a lesser or greater extent.

Some sensorial properties of synthetic sweeteners are a limiting factor for many applications; a combination of synthetic sweeteners into a mixture can overcome these limitations. Mixed sweeteners possess some novel attributes. For instance, there is an amplification, or synergy in their respective sweetness, usually in the range of 25 to 30 %, thus allowing for a further lowering of the dose of the total sweetener. Another benefit of the combined sweeteners is the possibility to bias the taste in order to achieve even a more pleasing sensation. Thus a combination of 1 part of the sodium salt of saccharin with 10 parts of cyclamate gives better, more natural sweetness and removes at the same time the unwanted aftertaste and off-taste [1]. The combinations of natural saccharides with synthetic sweeteners, or even mixtures of natural sugars, are also important. Thus synergy of between 10 to 50 % was found by Moskowitz [35] for nine different sugars at varying concentrations with glucose and fructose. However, difficulties were encountered by Moskowitz and Klarman [36] when studying the mixtures of saccharin and cyclamate. The sweetness was the prime quality at low concentrations, but at higher levels both sweetness and bitterness are exhibited by the same compound; therefore it was difficult to evaluate the component quality and intensity. Synergy was also noted in mixtures of Acesulfam K–sodium cyclamate and Acesulfam K–aspartame [2]. Synergism was also observed in mixtures of calcium cyclamate and dextrose or sucrose [3, 14]. It was also observed in mixtures of aspartame, hydroxypropyl cellulose, amylopectin and sodium chloride [24].

In order to assess the relative sweetness of synergistic mixtures of sweeteners, Hoppe and Gasmann [34] related the sweetness of 16 natural and synthetic sweeteners to their concentration (Table 4.1). The magnitude of sucrose sweetness at different concentrations is given together with the concentration of the synthetic sweetener required to achieve the same degree of sweetness. All data were obtained at room temperature. The solvent effect was negligible and was not taken into consideration. Temperature proved to be a factor in evaluating the sweetness of saccharin and cyclamate. The data in Table 4.1 permit the prediction of the sweetness of a combination of sweeteners, for example, how much saccharin must be added to a 50 g/l solution of sucrose in order that it matches the sweetness of an 80 g/l solution. Thus we need to add the sweetness potency of $80 - 50 = 30$ g/l of sucrose. This can be achieved by adding 51 mg/l of pure saccharin or $51.0 \times 1.175 = 59.92$ mg/l of commercial saccharin formulation. The coefficient 1.175 accounts for the effect of other compounds present in the mouth cavity that are responsible for the modification of the taste sensation, such as acid, organic and inorganic salts and other compounds that amplify or modify the taste. A mixture of Acesulfam K and sugar alcohols has, for example, a pleasant full taste which masks the original taste of the alcohols.

Many multiple sweeteners have been designed to mask the bitter aftertaste of saccharin and to accentuate its sweet taste. The combinations with fructose, glucose, starch hydrolysates, lactose, glycine, glutamic acid, glucono-1,4-lactone, glycols, citrates, sodium chloride, magnesium sulphate and potassium hydrogen sulphate are very typical [3, 5].

Commercial tablets of Acesulfam K contain an admixture of low viscosity carboxymethyl celluloses of polyvinylpyrrolidone. Effervescent tablets contain gelatin and sodium hydrogen carbonate. Powder formulations of Acesulfam K often contain also citrates, tartarates, sorbitol or lactose. Both aspartame and Acesulfam K combine well with sucrose, fructose and even xylose. The taste of Acesulfam K in admixture with multiple sweeteners achieves the top quality when the other component has the same sweetness magnitude.

Table 4.1 — Relation between concentration and degree of sweetness

Degree of sweetness	Sucrose	Maltose	Lactose	Glucose	Fructose	Galactose (g/l)	Xylose	Sorbitol	Mannitol	Xylitol	Lactitol	Palatinit	Saccharin (mg/l)	Cyclamate (g/l)	Acesulfam K (mg/l)	Aspartame (mg/l)
2	1.8	5.5	6.8	3.4	1.5	4.1	3.2	3.5	3.8	2.1	6.3	19.6	3.5	0.05	7.9	12.6
4	3.8	11.2	13.6	6.8	3.0	8.3	6.4	7.0	7.6	4.2	12.7	27.7	5.2	0.11	16.2	25.6
6	5.8	16.9	20.5	10.4	4.6	12.5	9.7	10.7	11.6	6.4	19.2	35.4	8.0	0.17	24.8	39.2
8	7.8	22.7	27.6	14.0	6.2	16.7	13.1	14.4	15.6	8.6	25.9	42.8	10.9	0.23	33.8	53.3
10	9.9	28.7	34.7	17.7	7.9	21.0	16.5	18.2	19.6	10.9	32.7	50.0	14.0	0.30	43.3	68.0
12	12.1	3.48	41.9	21.4	9.6	25.3	20.0	22.1	23.8	13.3	39.7	57.0	17.2	0.36	53.3	83.4
14	14.3	40.9	49.2	25.3	11.3	29.6	23.7	26.1	28.0	15.7	46.8	63.8	20.7	0.43	63.9	99.5
16	16.5	47.3	56.5	29.3	13.1	34.0	27.4	30.2	32.3	18.2	54.0	70.4	24.4	0.50	75.0	116
18	18.8	53.7	64.0	33.3	14.9	38.4	31.2	34.4	36.7	20.7	61.4	76.9	28.3	0.58	86.9	134
20	21.2	60.3	71.6	37.5	16.8	42.9	35.1	38.6	41.1	23.3	69.0	83.2	32.5	0.66	99.6	153
22	23.7	67.0	79.3	41.8	18.8	47.4	39.1	43.1	45.7	26.0	76.8	89.4	37.1	0.74	113	172
24	26.2	73.9	87.1	46.1	20.8	51.9	43.2	47.6	50.3	28.7	84.8	95.6	42.0	0.82	128	194
26	28.8	81.0	95.0	50.6	22.8	56.4	47.4	52.2	55.0	31.6	93.0	102	47.5	0.92	144	216
28	31.5	88.2	103	55.3	25.0	61.1	51.8	57.0	59.9	34.5	101	108	53.5	1.01	161	240
30	34.2	95.5	111	60.1	27.2	65.7	56.2	62.0	64.8	37.5	110	114	60.2	1.11	180	265
32	37.1	103	119	65.0	29.5	70.4	60.9	67.1	69.9	40.6	119	120	67.8	1.22	201	293
34	40.1	111	128	70.1	31.9	75.2	65.6	72.3	75.0	43.8	128	126	76.7	1.33	225	323
36	43.2	119	136	75.3	34.4	80.0	70.5	77.8	80.3	47.2	137	133	87.0	1.45	252	356
38	46.4	127	145	80.7	36.9	84.8	75.6	83.4	85.7	50.6	147	139	99.7	1.58	284	392
40	56.8	153	172	98.1	45.4	99.6	92.0	101	103	61.8	177	160	179	2.04	433	533

Table 4.1 (continued)

Degree of sweetness	Sucrose	Maltose	Lactose	Glucose	Fructose	Galactose (g/l)	Xylose	Sorbitol	Mannitol	Xylitol	Lactitol	Palatinit	Saccharin (mg/l)	Cyclamate (g/l)	Acesulfam K (mg/l)	Aspartame (mg/l)
42	60.6	162	181	104	48.4	105	97.9	108	109	65.8	188	167	—	2.23	530	596
44	64.6	171	190	111	51.6	110	104	115	115	70.1	200	174	—	2.44	739	673
46	68.8	181	200	118	55.0	115	110	122	121	74.5	211	182	—	2.67	—	772
48	49.7	135	154	86.3	39.6	89.7	80.9	89.2	91.2	54.2	157	146	116	1.72	322	433
50	53.2	144	162	92.1	42.4	94.6	86.4	95.2	96.9	57.9	167	153	139	1.88	369	479
52	73.1	191	209	125	58.6	120	117	129	128	79.1	223	190	—	2.94	—	913
54	77.8	201	219	132	62.4	125	124	137	134	84.0	236	198	—	3.26	—	—
56	82.7	212	229	140	66.4	130	131	145	141	89.1	249	207	—	3.65	—	—
58	87.9	223	239	148	70.7	136	139	153	148	94.6	263	216	—	4.16	—	—
60	93.4	234	250	156	75.3	141	147	162	155	100	277	226	—	4.87	—	—
62	99.4	246	260	165	80.2	147	156	172	—	106	292	236	—	6.09	—	—
64	106	259	271	175	85.6	152	165	182	—	113	307	246	—	—	—	—
66	113	272	282	185	91.5	—	174	192	—	120	324	257	—	—	—	—
68	120	—	293	196	97.9	—	185	204	—	128	341	268	—	—	—	—
70	129	—	304	208	105	—	196	216	—	136	359	280	—	—	—	—
72	138	—	316	220	113	—	208	229	—	145	337	293	—	—	—	—
74	148	—	328	234	122	—	221	244	—	155	399	306	—	—	—	—
76	160	—	—	248	133	—	235	260	—	166	420	320	—	—	—	—
78	174	—	—	265	146	—	251	277	—	179	444	—	—	—	—	—
80	191	—	—	282	162	—	269	296	—	194	469	—	—	—	—	—

Table 4.1 (continued)

Degree of sweetness	Sucrose	Maltose	Lactose	Glucose	Fructose	Galactose (g/l)	Xylose	Sorbitol	Mannitol	Xylitol	Lactitol	Palatinit	Saccharin (mg/l)	Cyclamate (g/l)	Acesulfam K (mg/l)	Aspartame (mg/l)
82	212	—	—	303	184	—	289	318	—	211	496	—	—	—	—	—
84	240	—	—	326	217	—	312	344	—	232	526	—	—	—	—	—
86	282	—	—	353	287	—	339	374	—	259	559	—	—	—	—	—
88	374	—	—	386	—	—	373	411	—	298	595	—	—	—	—	—
90	—	—	—	427	—	—	416	459	—	364	636	—	—	—	—	—
92	—	—	—	481	—	—	477	526	—	—	683	—	—	—	—	—
94	—	—	—	564	—	—	581	640	—	—	737	—	—	—	—	—
96	—	—	—	—	—	—	—	—	—	—	803	—	—	—	—	—
98	—	—	—	—	—	—	—	—	—	—	885	—	—	—	—	—
100	—	—	—	—	—	—	—	—	—	—	995	—	—	—	—	—

4.1 MIXTURES OF NATURAL SWEET SUBSTANCES

Most natural mixed sweeteners attempt to modify the properties of sucrose. The addition of cheap, but highly potent sugars such as fructose (in the ratio 1 : 1.5 or 1.5 : 1) [12] or of the 1 : 1 fructose–glucose syrup can increase the sweetness on the weight basis [4].

The sucrose–sorbitol (or other sugar alcohols) combination has been used in chewing gums. Starch hydrolysates are often combined not only with sucrose but also with sugar alcohols and malic acid in order to increase the flavour. Sometimes sodium phosphate and maltol are also added.

The relative sweetness of blends of natural sugars (sucrose *vs.* fructose, lactose, glucose, galactose, mannitol, glycine and glycerine) was determined by Porter [1].

A pleasing sweet taste can be imparted to soft drinks by a blend of neohesperidin dihydrochalcone and sucrose 1 : 1 [6, 7]. The attractive natural sweet taste of sucrose has often been amplified by maltol, ethyl maltol, thaumatin or talin [6].

A patent has been issued for a non-carcinogenic dry mix sweetener consisting of a mixture of amino acids — glycine, alanine, β-alanine, tryptophan, hydroxyproline and proline [13].

4.2 SYNTHETIC MULTIPLE SWEETENERS

The more common components that we encounter in the blends of synthetic sweeteners are saccharin, cyclamates and Acesulfam K; less common ones are aspartame and others.

Dry mixes for non-alcoholic beverages are often a mixture of saccharin and sorbitol in concentrations corresponding to one gram of sucrose (Sionon, Gluconon) [6]. Xylitol–fructose blends lack the aftertaste of saccharin and perform well in lemonades as well as in synergism tests at temperatures between 5, 23 and 50 °C [8]. Different blends of aspartame and saccharin (1 : 15, 15 : 1, 48 : 1) have also been tried in dry mix beverages [5]. Cola drinks are sweetened by sodium saccharinate–aspartame (1 : 1) or by sodium saccharinate–cyclamate (1 : 10) mixtures. Calcium sulphamate present in soft drinks in 0.75 to 0.86 % concentrations, or aspartame in 0.17 to 0.19 % concentration, are equivalent to a 10 % solution of sucrose [15]. A mixture that is stable in acidic media at pH < 4, and utilized in orange juice, was prepared from aspartame and glycosyl stevioside (1–10 : 1) [27].

Food and drink acquire natural and pleasant sweetness by an addition of 0.03 % of a blended sweetening agent that consists of 1–2 : 6–1 ratio of aspartame–cyclamate [28]. The 1 : 1 mixture of Acesulfam K–aspartame suits drinks better; both sweeteners are present in 200–300 mg/l concentrations. The blended sweetener has better stability than when aspartame is used alone; the combination also improves the quality of flavour of the resulting sweet taste. Interesting variations in taste can be achieved by varying the relative concentrations of cyclamate and Acesulfam K [2]. When the sodium cyclamate content reaches 900 mg/l, further 200–300 mg/l of Acesulfam K can be added.

An addition of small amounts of fructose to synthetic sweeteners often helps to improve the taste deficiencies of artificial sweeteners. Using parallel comparison tests, non-linear sweetness *vs.* concentration relationships have been established for fructuse, saccharin, aspartame and Acesulfam K against sucrose in acidified non-carbonated mineral water [30]. Utilizing the high relative sweetness of fructose, low-calorie soft drinks containing as little as 2–3 % of sugar could not be distinguished by taste from the traditional sucrose drinks.

Non-alcoholic beverages in Japan are sweetened by fructose blend with glycosyl stevioside [5] or with a combination of neohesperidin dihydrochalcone and saccharin or cyclamates (11, 25 or 64 %). The sweetness is magnified in such mixtures and the flavour is improved.

The favoured powder sweetening mixtures used for drinks have been saccharin–sucrose blends [9], as well as saccharin–cyclamate (1 : 10) mixture and thaumatine or monellin with an addition of arabic gum (1 : 9) [32]. A sweetener has been designed for use in coffee which consists of a mixture of Acesulfam K, glucose and potassium tartarate in the ratio of 5 : 95 : 2 [29]. A mixture of 50 g sorbose, 1 g aspartame and 50 ml water can serve as a useful dry-mix sweetener [20].

Tablets of tabletop sweeteners have been prepared by mixing sodium saccharinate and Dulcin [3], sodium saccharinate and aspartame (4: 2 mg), sodium saccharinate, aspartame and sodium cyclamate (4 : 10 : 30 mg) [6] or sodium saccharinate and cyclamate (1 : 2, 1 : 3, 1 : 10 or 1 : 12). A mixture of saccharin and cyclamate (1 : 10) has a relative sweetness of about 63, which corresponds to a solution of 100 g sucrose per litre. Moreover, the mixture has a clean taste, without any aftertaste.

At higher levels of cyclamate and saccharin, each compound exhibits its own off-taste apart from the sweetness. Being aware of this, Vincent and co-workers [37] at Abbot Laboratories determined the level of cyclamate needed in order to reduce the incidence of the off-taste to only 20 % of their panel of tasters. They then combined this with saccharin in a 10 : 1 ratio (cyclamate–saccharin) and found much improved sweetness with reduced aftertaste and 50–100 % synergy.

The cyclamate ban in the US in 1970, which was closely followed by other countries, reduced significantly the ability of the manufacturers to formulate products with higher consumer acceptance. Tabletop products made solely with one of the two sweetners are significantly *inferior* to products that are formulated as mixed sweeteners. This point is illustrated in Fig. 4.1 which compares the sweetness profile of sucrose with a 10 : 1 cyclamate–saccharin mixture.

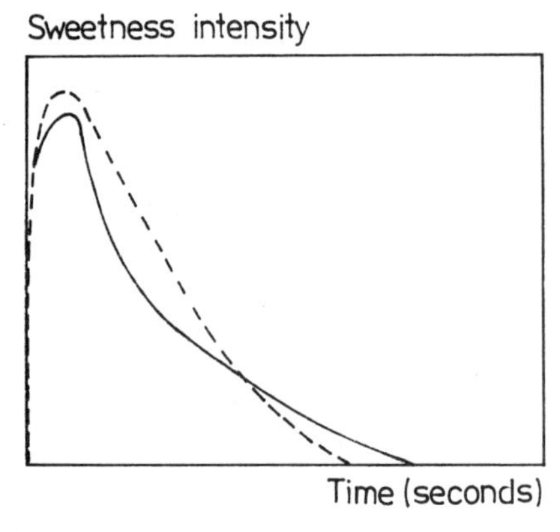

Fig. 4.1 — Sweet potency of sucrose (— — —), and of mixture of sodium saccharin–cyclamate 1 : 10 (———).

The elimination of saccharin aftertaste has been demonstrated also for tablets containing sodium cyclamate, sorbitol and glucose [10]. The tabletop sweetener tablets (Polisette) have been marketed in Hungary; they contain a mixture of 19 parts by weight of sodium cyclamate and one part by weight of sodium saccharinate [6]. Another formulation of sweetener tablets is a ternary mixture of aspartame, lactose and dextrin (5 : 87 : 8 by weight), 5 times sweeter than sucrose [25]. Effervescent and highly soluble 1 g tablets contain *N*-acetylcysteine (6–32 parts by weight), citric acid (36–60), $NaHCO_3$ (26–37), aspartame (1–1.5) and aromatic flavouring (5–7) [26].

When a decreased calorie content is desired, saccharin can be combined with other less intense sweeteners such as xylose, glucose syrup or a mixture of fructose, xylitol and cyclamate [6]. A special mixture has been designed for diabetics; it contains xylitol (44.84 g), sorbitol (179 g) and aspartame (0.35 g). It imitates very closely the sweet taste of sucrose [19].

A mixture consisting of 3 mg of sodium saccharinate and 30 mg of sodium cyclamate in 800 g of a carrier was used for sweetening cocktails, puddings and pastries [5].

Cyclamate–saccharin combinations were used extensively in baked goods because of the thermal stability of the mixture. The ratio of cyclamate to saccharin varied from 2 : 1 to 10 : 1, the latter being the preferred ratio when organoleptic considerations were taken into account.

Chewing gums are another application area for saccharin–cyclamate mixtures which now typically contain 1 : 2 or 1 : 10 ratio of the components. Sodium saccharinate–aspartame in the ratio of 1 : 2 have been also used for this purpose [5]. Yet another combination that has been recently patented is 99.5–99.0 : 0.5–1 per cent by weight ratio of sodium saccharinate and either thaumatine or monellin [11].

The utilization of a fructose–sodium cyclamate–maltol mixture in canned food and in the processing of paprika to a ground condiment have been reported [6]. Also, a patent has been issued for the use of a glucose–fructose–aspartame mixture in fermented dairy products [23].

Aspartame has been also combined with sugar alcohols and sodium tartarate [17], with sorbitol in 20–0.1 : 80–99.9 per cent by weight [18], as well as with glucose, branched oligosaccharides or starch [21, 22]. Some other combinations are also described [16].

Combinations of synthetic sweeteners with α-aminocarboxylic acids have been suggested, and a mixture of Acesulfam K with glycine, histidine, arginine, glutamine, tartarates and sodium hydrogen phosphate was patented [31].

Since the clarity of sweet taste in food applications is considered of utmost importance, a suggestion has been made that the average values of sweet potency of sodium saccharinate, Acesulfam K, cyclamates and aspartame be considered as the upper limit at which the taste purity is preserved [33].

It has already been stressed that the combination sweeteners have decided advantages over pure ones, therefore it should not be surprising that novel combinations and formulations are being published, and it is next to impossible to trace them all.

REFERENCES

1. Porter, A. B. (1983) *Chem. Ind.* 696.
2. von Rymon Lipinski, G. W. (1986) *Alimenta* **25**, 150.
3. Tapodo, J. (1977) *Édesipar* **28**, 2.
4. Pohlová, M. (1981) *Prům. Potr.* **32**, 416 (in Czech).
5. Bakal, I. A. (1983) *Chem. Ind.* 700.

6. Vukov, K. (1986) *Cukoripar* **39**, 72.
7. Beerens, H. (1981) *Ann. Fals. Exp. Chim.* **74**, 261.
8. Hyvönen, L., Kurkela, R., Koivistoinen, P. & Patilainen, A. (1978) *J. Food Sci.* **43**, 251.
9. Ger. Offen. 3329764 (1983) [*Chem. Abstr.* **102**, 1985, 202903].
10. Brit. Pat. Appl. 2187074 (1987) [*Chem. Abstr.* **107**, 1987, 216458].
11. U.S. 758438 (1988) [*Chem. Abstr.* **109**, 1988, 229165].
12. U.S. 4676991 (1987) [*Chem. Abstr.* **107**, 1987, 95616].
13. Eur. Pat. Appl. 65462 [*Chem. Abstr.* **98**, 1983, 106001].
14. Stone, H. & Oliver, S. M. (1969) *J. Food. Sci.* **34**, 215.
15. Hoppe, K. (1981) *Nahrung* **25**, 769.
16. Homler, B. E. (1984) In: Stegink, L. D. & Filler, L. J. (eds) *Aspartame: Physiology and Biochemistry*. Marcel Dekker, New York, p. 247.
17. Japan 62195265 [*Chem. Abstr.* **108**, 1988, 4923].
18. Ger. Offen. 3541302 (1987) [*Chem. Abstr.* **107**, 1987, 76451].
19. Swiss 646843 (1984) [*Chem. Abstr.* **102**, 1985, 111888].
20. Japan 6119466 [*Chem. Abstr.* **104**, 1986, 205828].
21. Japan 6011467 and 60114168 [*Chem. Abstr.* **104**, 1986, 4840, 4841].
22. Japan 61224963 [*Chem. Abstr.* **106**, 1987, 48962].
23. Japan 58175436 [*Chem. Abstr.* **100**, 1984, 50254].
24. Czech. 195781 (1982) [*Chem. Abstr.* **98**, 1982, 3754].
25. Eur. Pat. Appl. 106910 (1984) [*Chem. Abstr.* **101**, 1984, 169409].
26. Ger. Offen 3723734 (1988) [*Chem. Abstr.* **108**, 1988, 226873].
27. Japan 60221056 [*Chem. Abstr.* **104**, 1986, 128631].
28. Japan 6066954 [*Chem. Abstr.* **103**, 1985, 140651].
29. Ger. Offen. 3309426 [*Chem. Abstr.* **99**, 1983, 193547].
30. van Tornout, P., Pelgroms, J. & van der Meeren, J. (1985) *J. Food Sci.* **50**, 469.
31. Japan 59154957 [*Chem. Abstr.* **102**, 23197 (1985)].
32. Brit. Pat. Appl. 2123672 (1983) [*Chem. Abstr.* **101**, 1984, 22240].
33. Hoppe, K. & Gasmann, B. (1985) *Nahrung* **29**, 417.
34. Hoppe, K. & Gasmann, B. (1985) *Lebensmittelindustrie* **32**, 227.
35. Moskowitz, H. R. (1973) *J. Exp. Psychol.* **99**, 88.
36. Moskowitz, H. R. & Klarman, L. (1975) *Chem. Sense Flavour* **1**, 411.
37. Vincent, H. C., Lynch, M. J., Pohley, F. M., Helgren, F. J. & Kirchmeyer, F. J. (1955) *J. Amer. Pharm. Soc.* **XLIV**, 442.

Subject index